User Guide

Picture Tests

The tests in this book are based on photographs of dissections published in the authors' *Human Anatomy: text and colour atlas Second Edition*. Each test comprises one or two photographs. You are required to identify the area illustrated, name all the structures indicated and then answer the questions presented.

Captions

If you have difficulty identifying the part of the body illustrated by a photograph, turn to the captions which provide a basic description of each dissection but do not answer the questions.

Answers

Answers comprise a full list of the names of the structures indicated, plus short statements satisfying the questions. In addition, the numbers of relevant pages in *Human Anatomy: text and colour atlas, Second Edition* are indicated so that you can refer to the anatomy on which each test is based.

Contents

HUMAN

PICTURE TESTS

anatomy

J. A. Gosling MD, MB ChB
Professor of Anatomy
Chinese University of Hong Kong
Hong Kong

J. R. Humpherson MB ChB
Senior Lecturer in Anatomy
Department of Cell
    and Structural Biology
University of Manchester, UK

P. L. T. Willan MB ChB, FRCS
Senior Lecturer in Anatomy
Department of Cell
    and Structural Biology
University of Manchester, UK

P. F. Harris MD, MB ChB, MSc
Professor and
Head of Department
of Human and
Clinical Anatomy
Sultan Qaboos University
Muscat, Sultanate of Oman

I. Whitmore MD, MB BS, LRCP, MRCS
Senior Lecturer in Anatomy
Department of Anatomy
Queen Mary and
Westfield College
London, UK

Photography by:
A. L .Bentley ABIPP, AIMBI, MBKS
Medical Photographer
Department of Cell
    and Structural Biology
University of Manchester, UK

J. L.Hargreaves BA (Hons)
Medical Photographer
Department of Cell
    and Structural Biology
University of Manchester, UK

Gower Medical Publishing

LONDON • NEW YORK

Distributed in the USA
and Canada by:
Raven Press Ltd.
1185 Avenue of the Americas
New York
New York 10036
USA

Distributed in the rest of the world by:
Gower Medical Publishing
Middlesex house
34-42 Cleveland Street
London W1P 5FB
UK

Library of Congress Cataloging in Publication Data and British Library
Cataloguing in Publication Data are available upon request.

ISBN 1-563-75542-4

Publisher: Fiona Foley

Project Manager: David Cooke

Design and Layout: Peter W. Wilder & Babes Koura

Paste Up: Ruth S.Miles & Olgun Hassan

Production: Susan Bishop
Adam Phillips

Originated in Hong Kong by Mandarin Offset (HK) Ltd

Typesetting by PCS, Frome

Produced by Mandarin Offset. Printed in Hong Kong.

PICTURE
TESTS

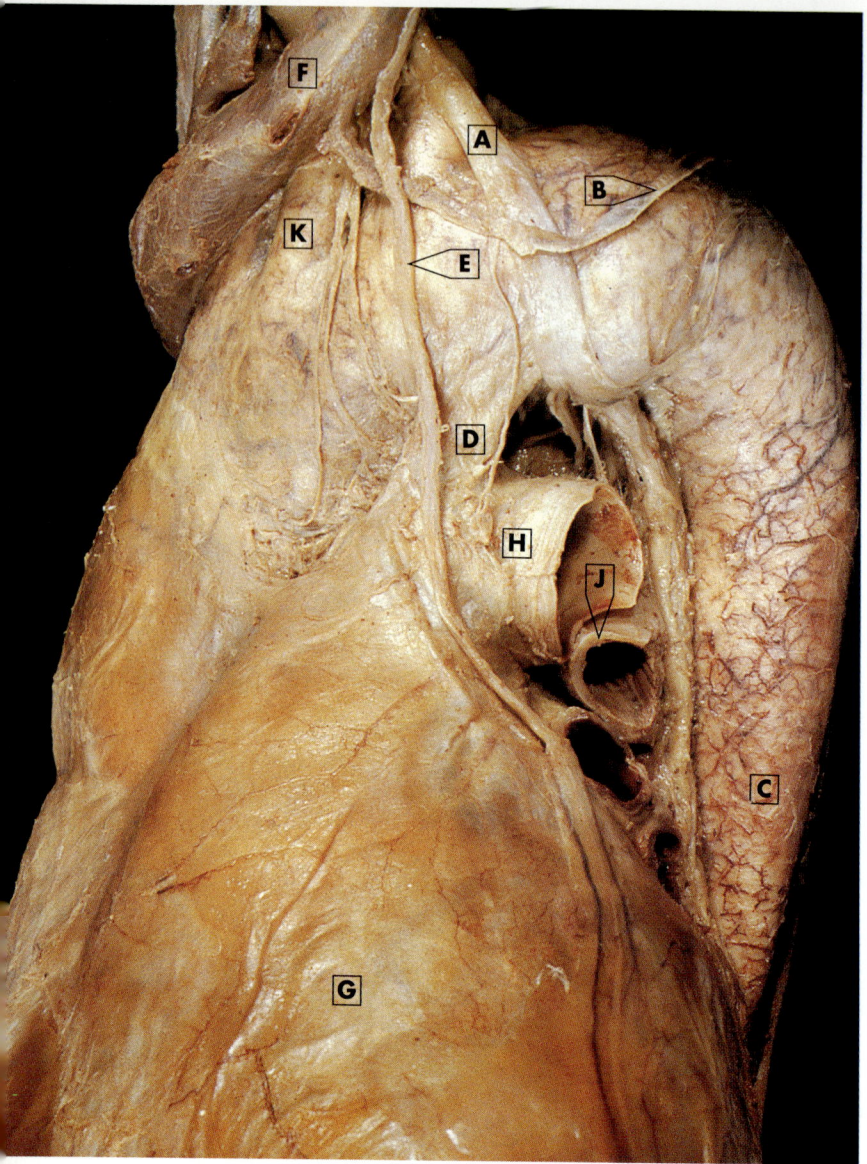

1. Which intercostal spaces are drained by B?
2. What is the vertebral level at which C enters the abdomen?
3. What is the embryological significance of D?
4. What is the innervation of G?
5. What are the branches of J?

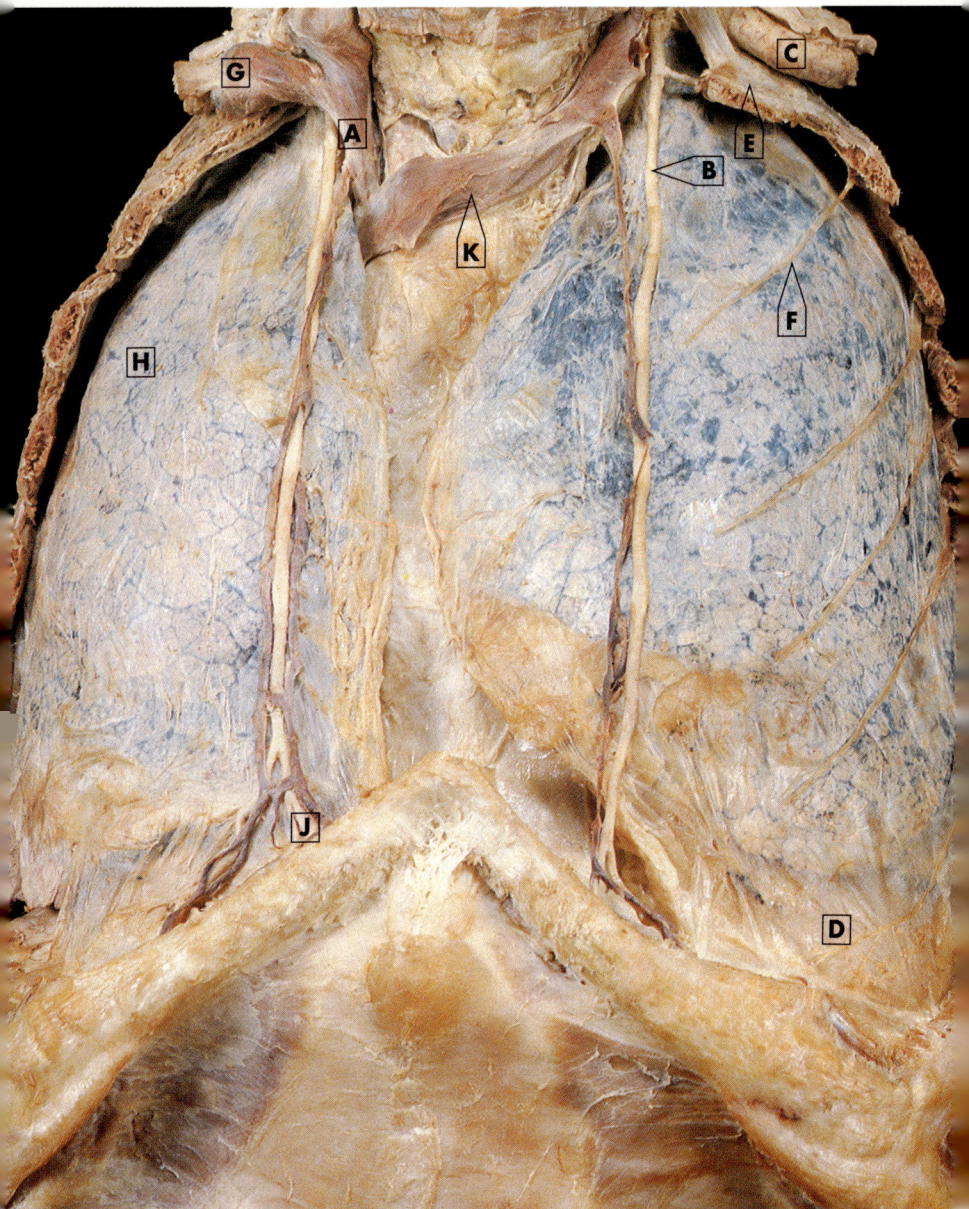

1. What is the origin of B?
2. What is the name of the pleural recess D?
3. What is the origin of G?
4. What is the innervation of H?
5. What is the part of the mediastinum which contains K?

3

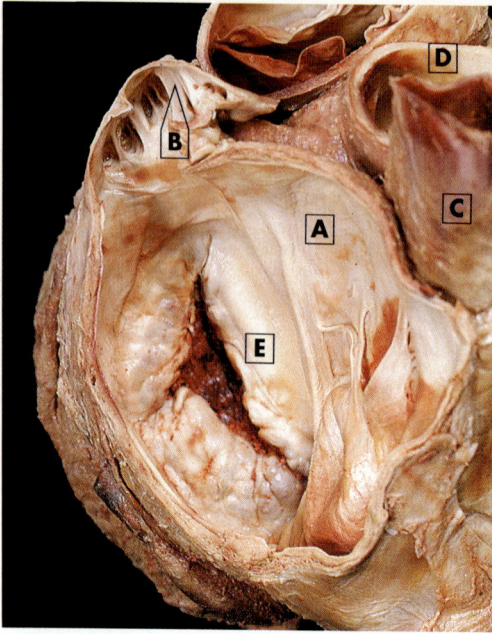

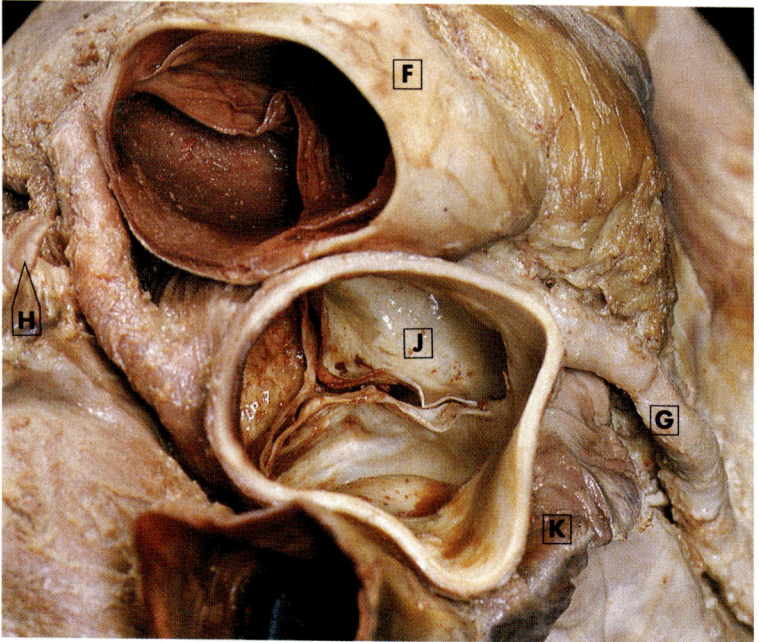

1. What are the muscular ridges in B?
2. What attaches to the free edge of E?
3. What is the first named branch of G?
4. What is the function of J?

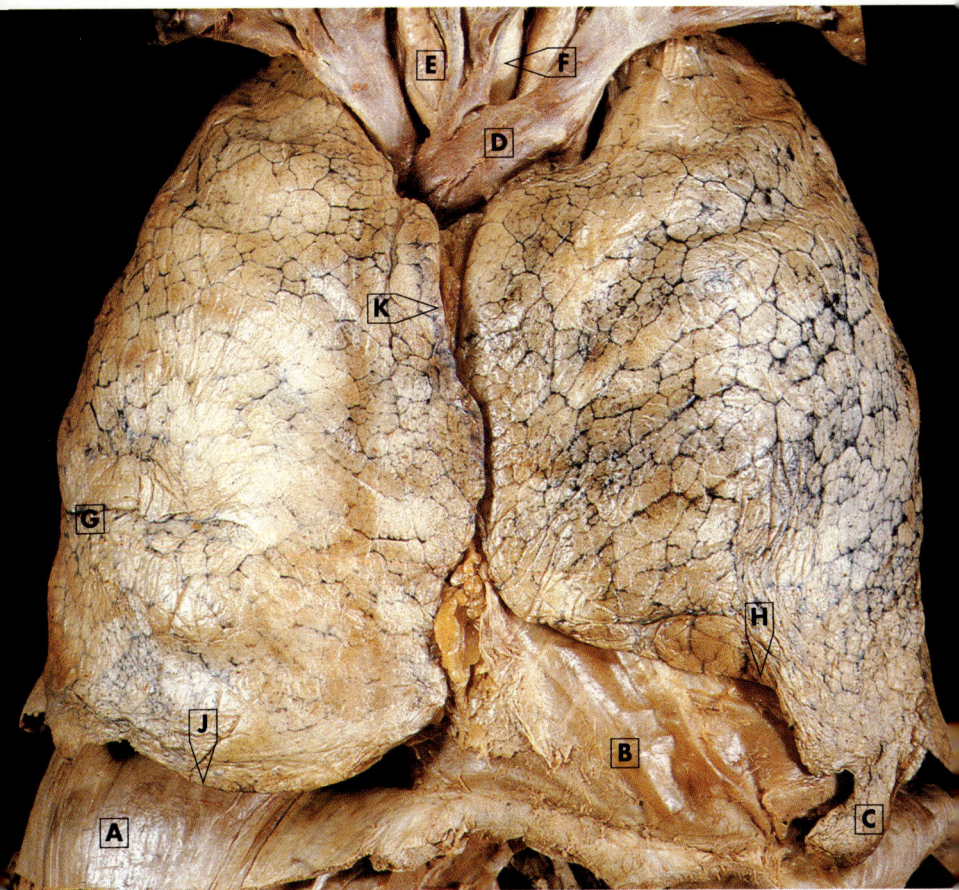

1. What is the nerve supply of the central part of A?
2. What is the nerve supply of B?
3. What is the origin of D?
4. What is the origin of E?

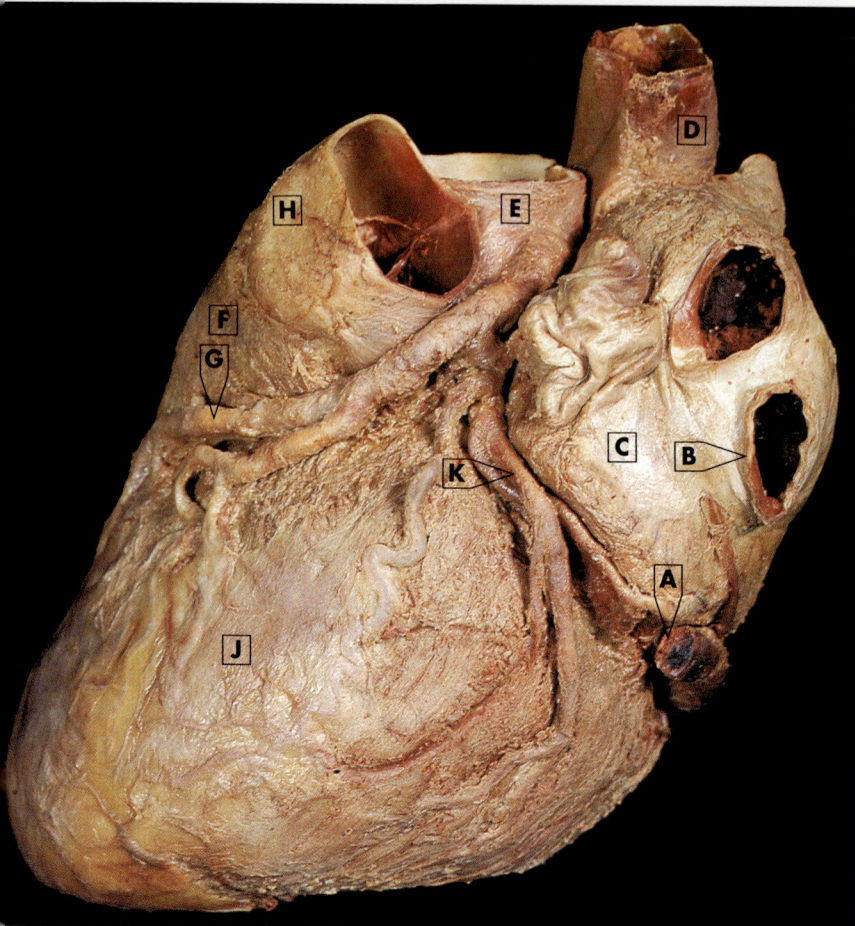

1. What is the termination of A?
2. What are the tributaries of A?
3. What part of the mediastinum contains C?
4. What part of the mediastinum contains E?
5. What is the name of this part of chamber F?
6. What is the termination of K?

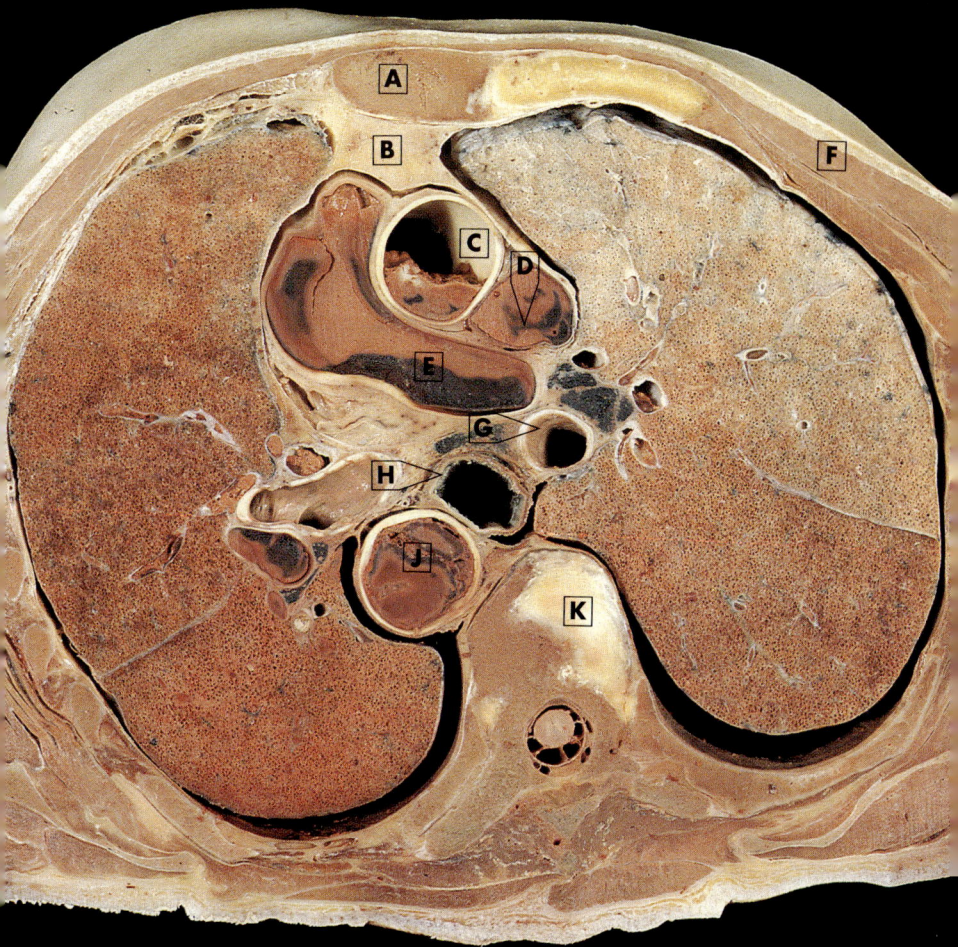

1. What tissues form B?
2. What is the continuation of C?
3. What is the origin of E?
4. Which heart chamber is related to H?
5. What is the continuation of J?

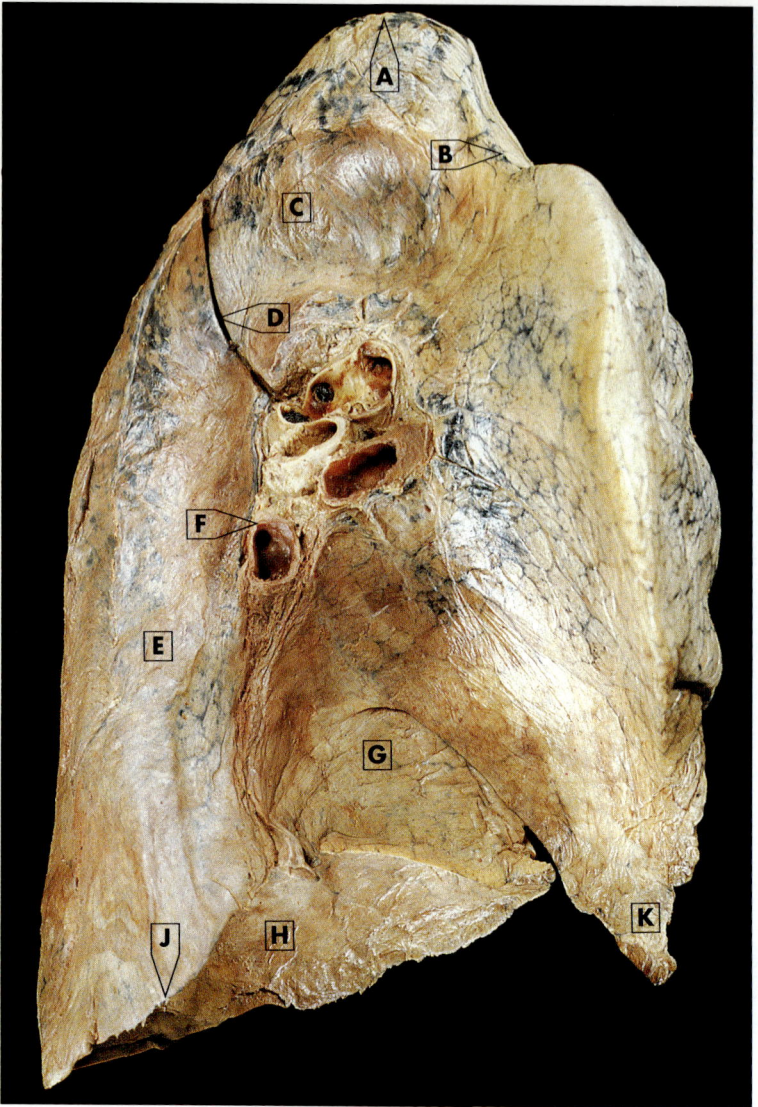

1. What is the surface marking of A?
2. Where does the structure terminate which caused notch B?
3. What are the branches of the structure which caused groove C?
4. What pleural fold is related to F?
5. Which heart chamber was related to G?
6. Which is the pleural recess related to J?

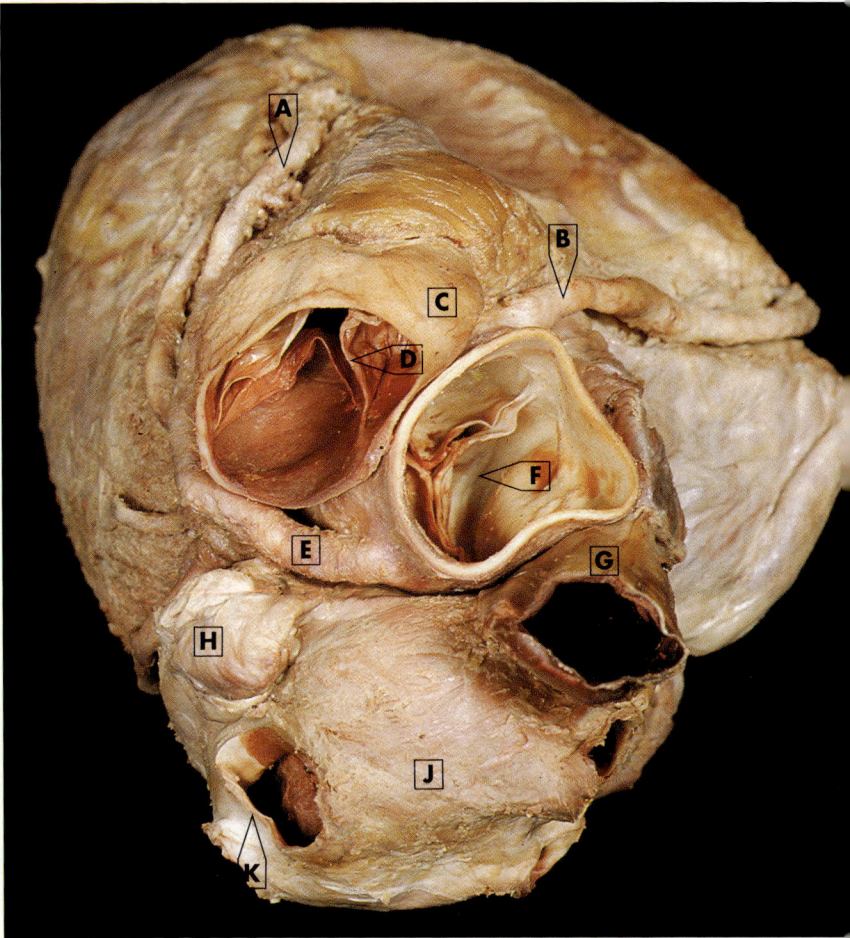

1. What is the swelling at C?
2. Where is the location of the loudest sounds caused by D?
3. Where is the location of the loudest sounds caused by F?
4. What is the continuation of E?
5. What is the valve guarding chamber J?

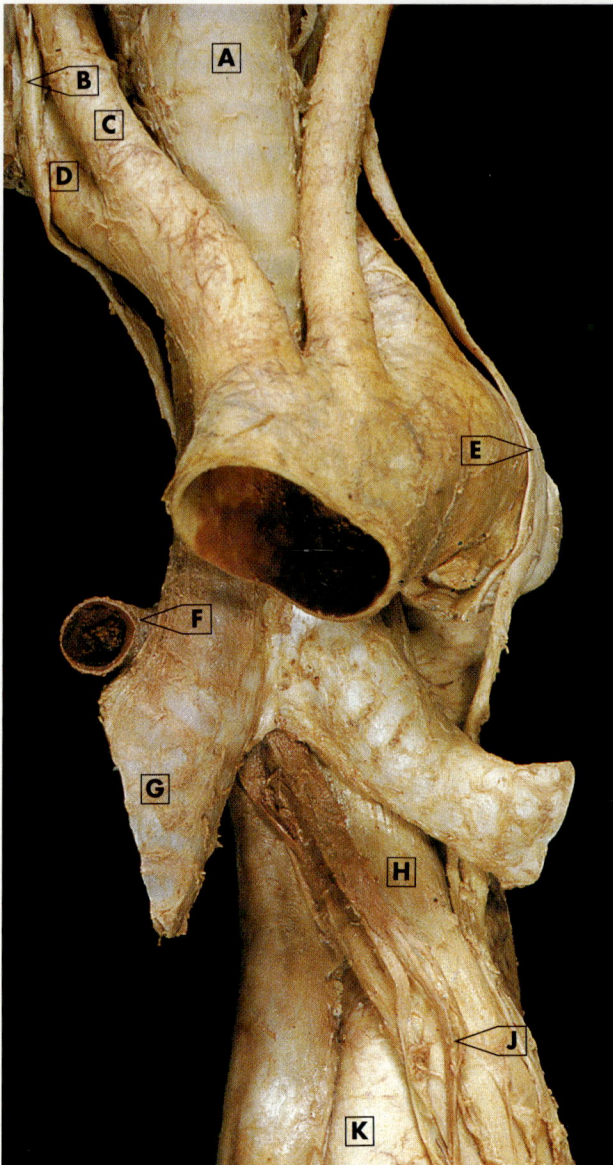

1. What is the vertebral level of the termination of A?
2. What is the branch of B which is related to D?
3. What is the termination of F?
4. What is the vertebral level of the diaphragmatic opening for H?
5. What is the vertebral level of the diaphagmatic opening for K?

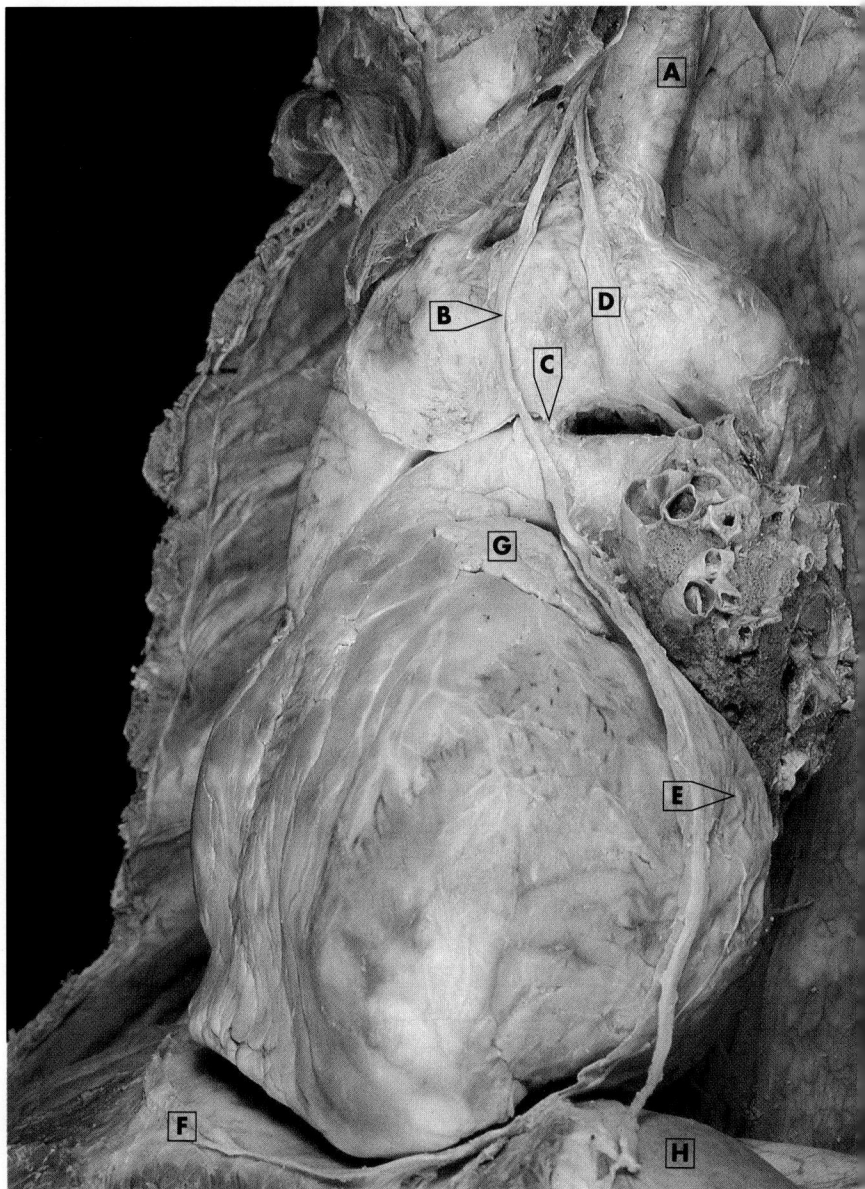

1. What is the termination of A?
2. What is the motor distribution of B?
3. What branch of D is related to C?
4. What is the blood supply of E?
5. What is the sensory nerve supply of F?
6. What is related to the abdominal surface of F?

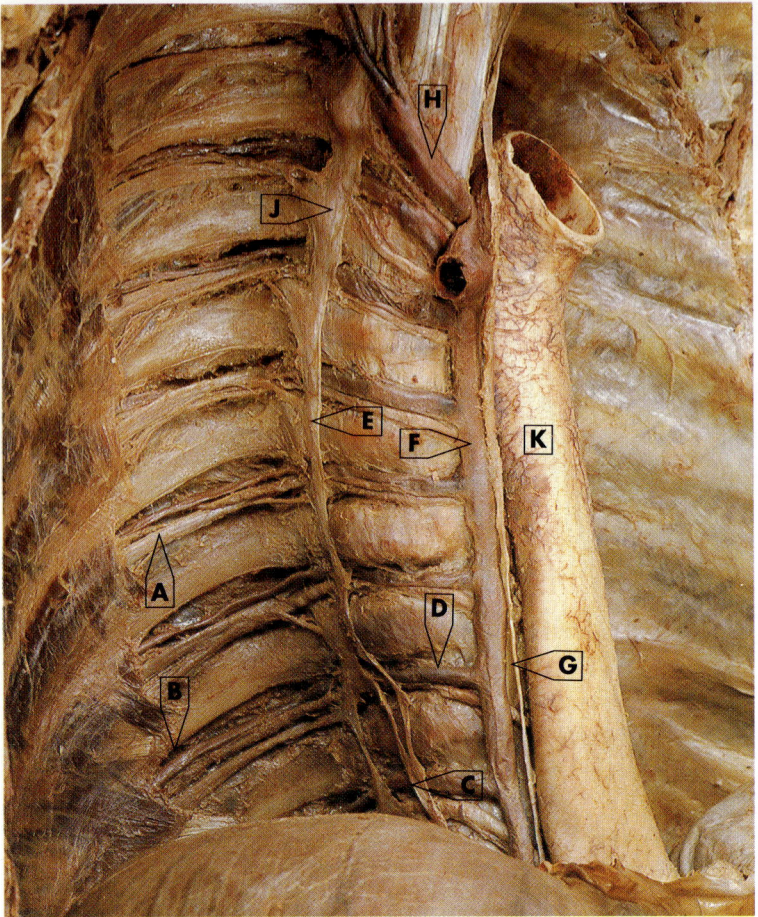

1. What muscles are supplied by A?
2. What part of the diaphragm is traversed by C?
3. What part of the diaphragm is related to E?
4. What part of the diaphragm is traversed by F?
5. What is the termination of G?

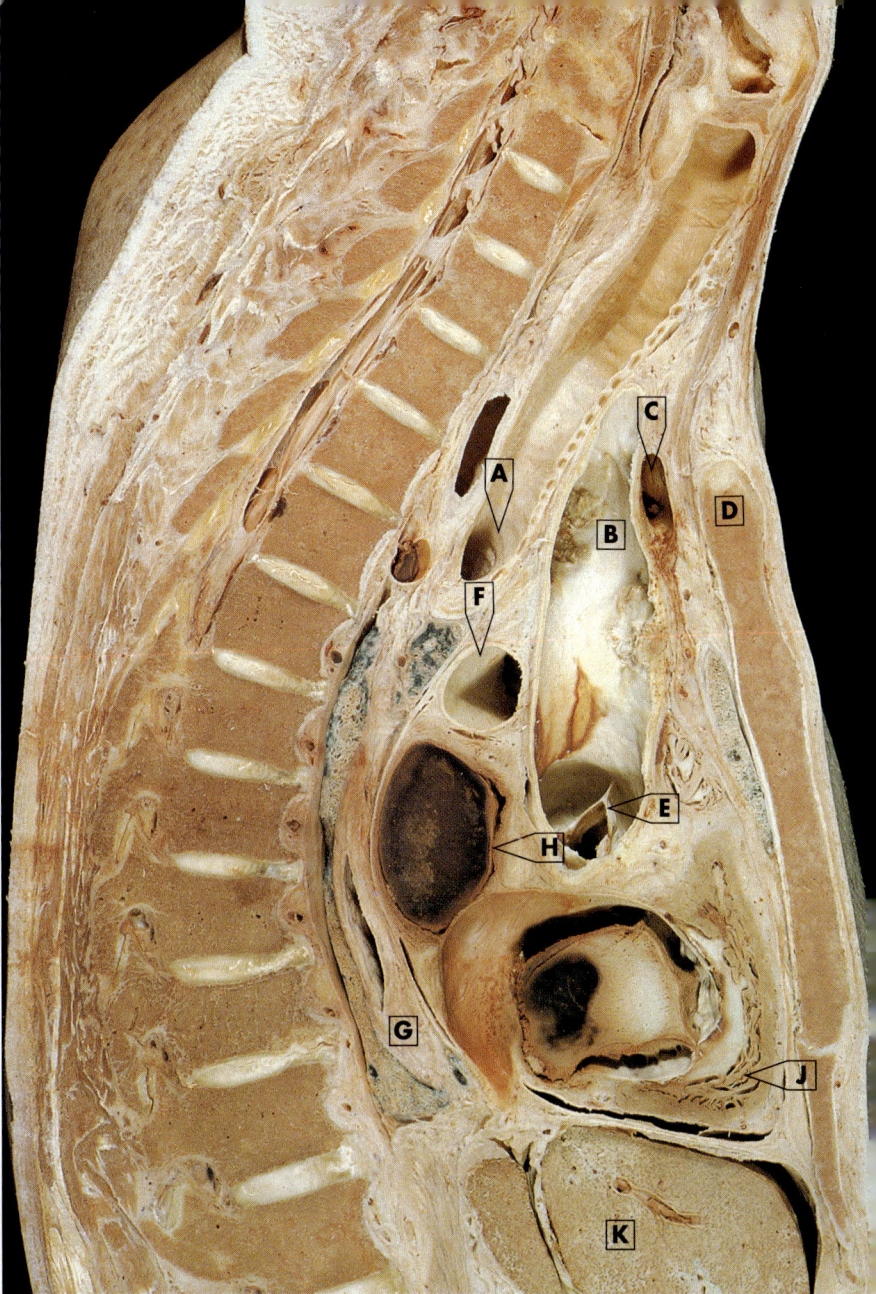

1. What part of the mediastinum contains B?
2. What is the termination of C?
3. What muscles attach to the posterior surface of D?
4. What is the vertebral level of the diaphragmatic opening for G?
5. What valves guard chamber J?

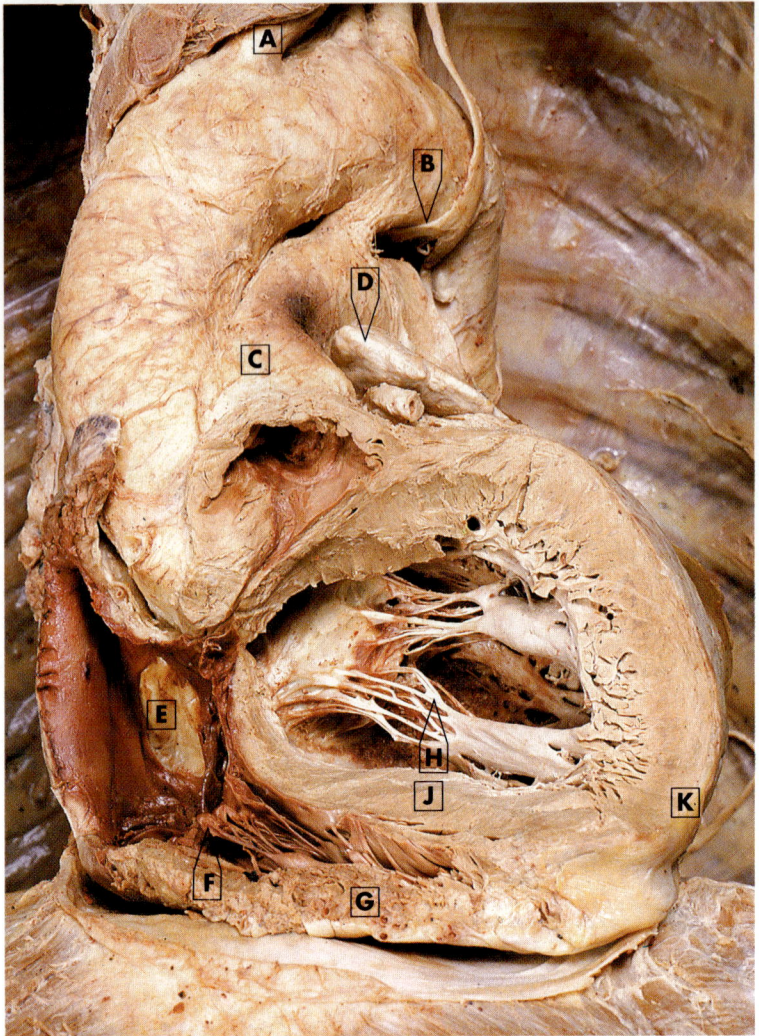

1. What is the embryological significance of E?
2. What is the location of the loudest sounds produced by F?
3. What is the main blood supply of G?
4. What is the main blood supply of J?
5. What is the surface marking of K?

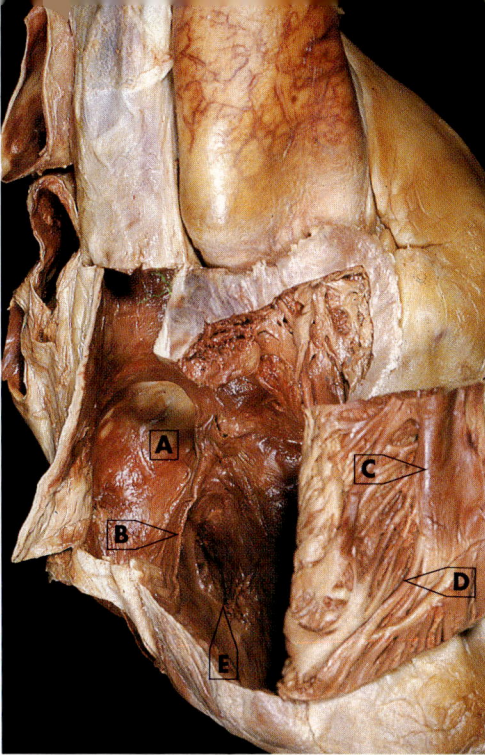

1. What is the embryological significance of B?
2. What lies at the superior end of C?
3. What is the origin of G?
4. What is the nerve supply of K?

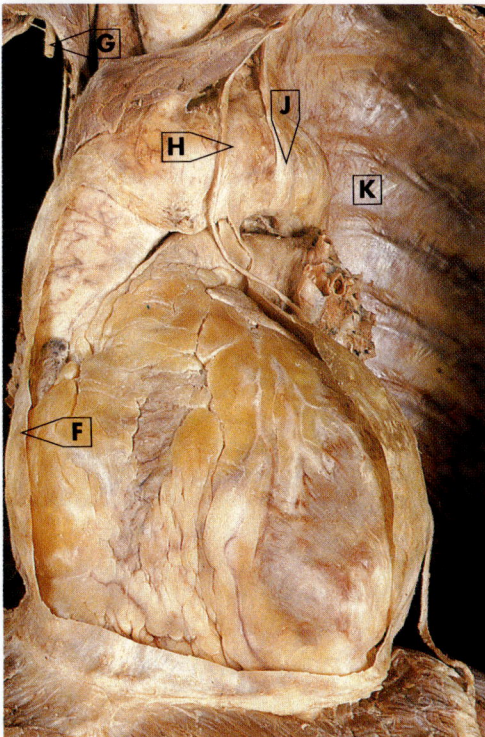

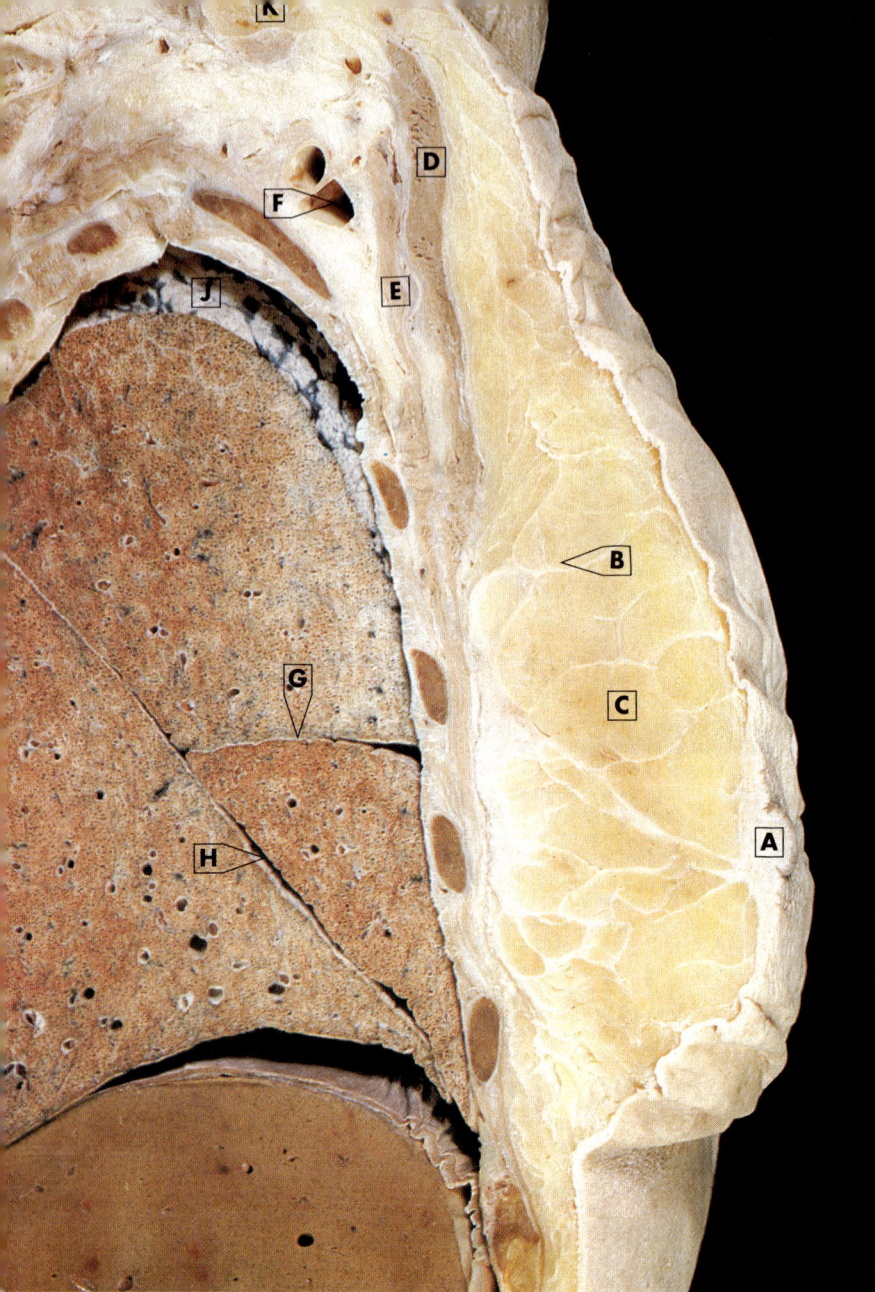

1. What structures terminate on the surface of A?
2. What is the nerve supply of E?
3. What are the actions of E?
4. What is the origin of F?
5. What lies between G and H?

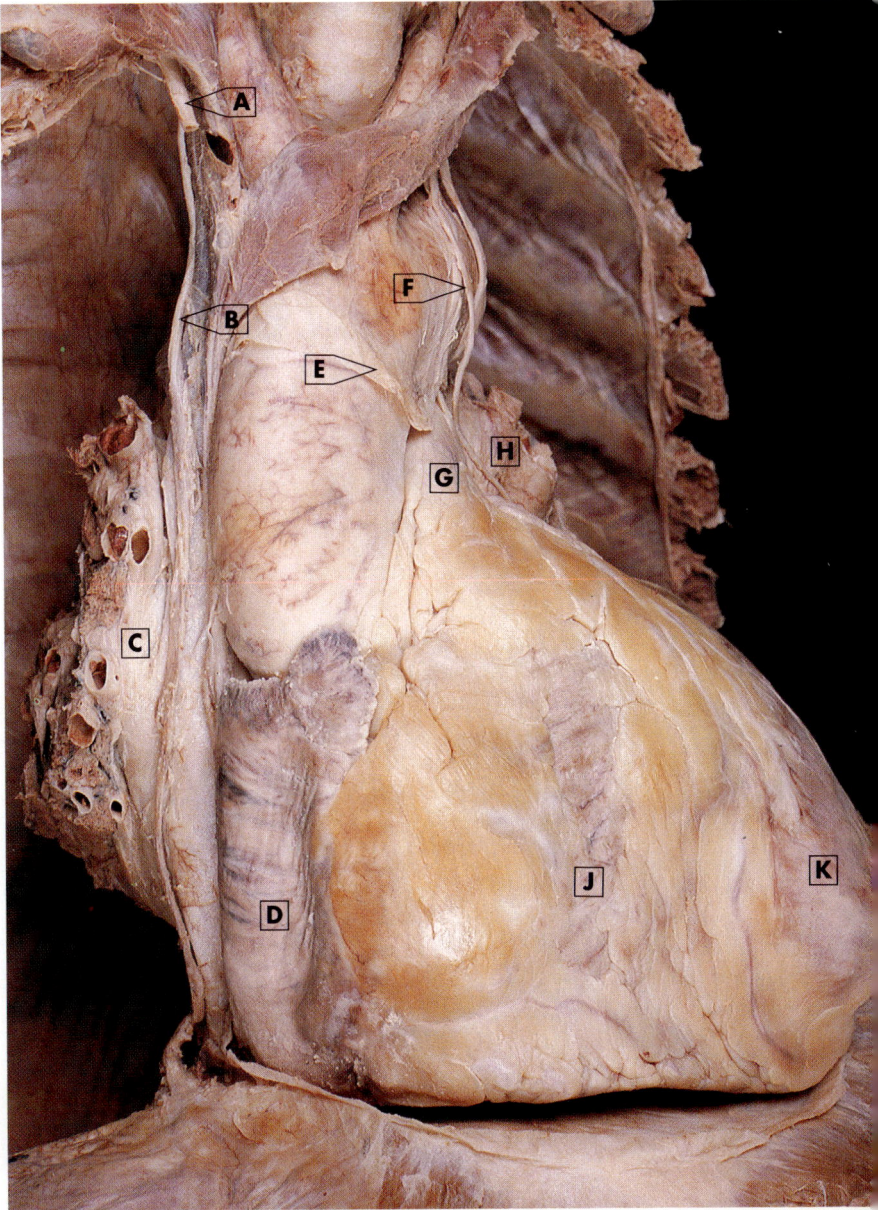

1. What are the principal components of C?
2. What vessels drain into D?
3. What is the inferior attachment of E?
4. What is the origin of G?
5. What valves guard J?

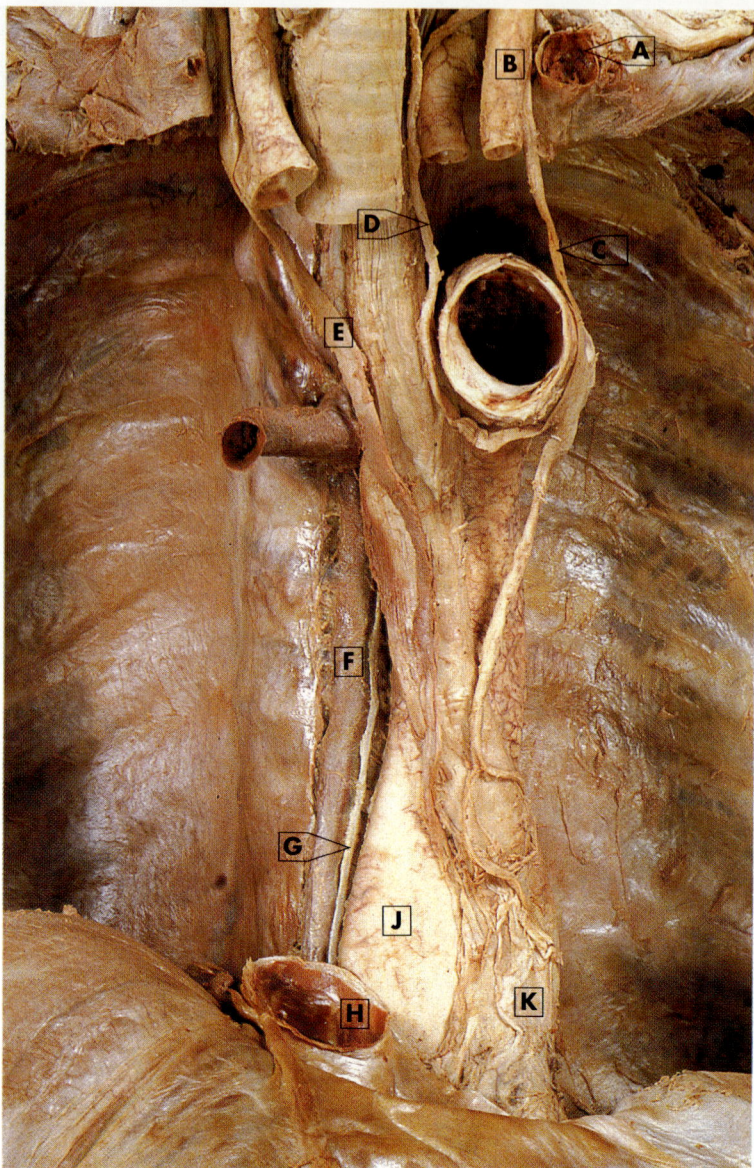

1. What part of the mediastinum contains B?
2. What opening in the diaphragm is transversed by F?
3. What opening in the diaphragm is transversed by G?
4. What is the vertebral level of the opening in the diaphragm for H?
5. What passes through the diaphragm alongside H?
6. What is the venous drainage of K?

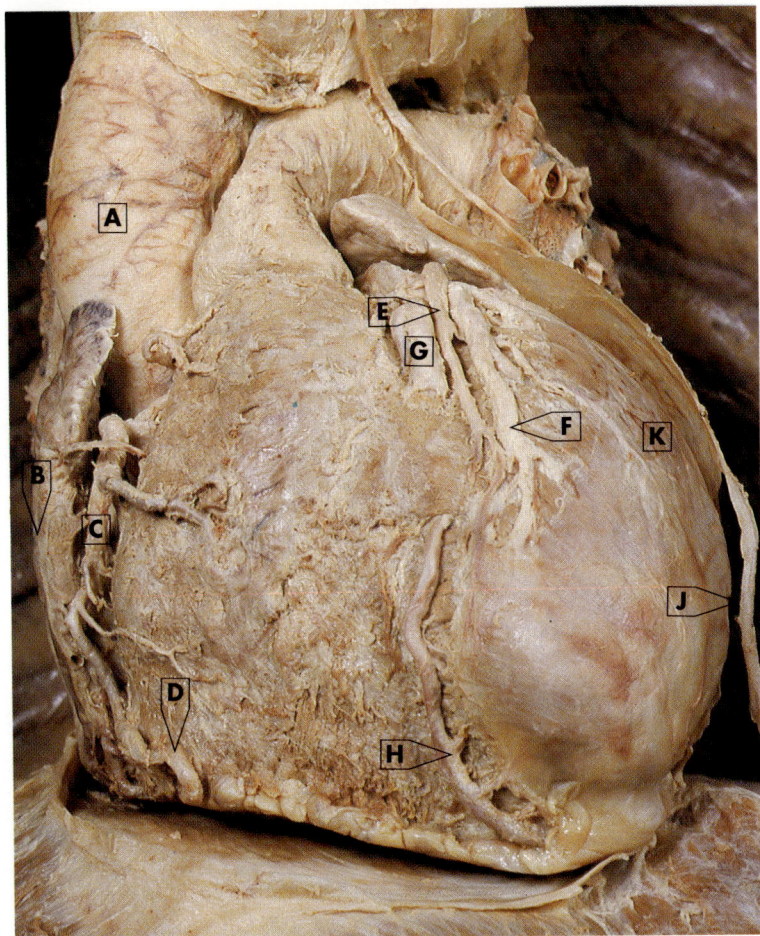

1. What are the surface markings of B?
2. Where is the termination of D?
3. What is the termination of E?
4. Where is the location of referred pain carried by J?
5. What are the surface markings of the border at K?

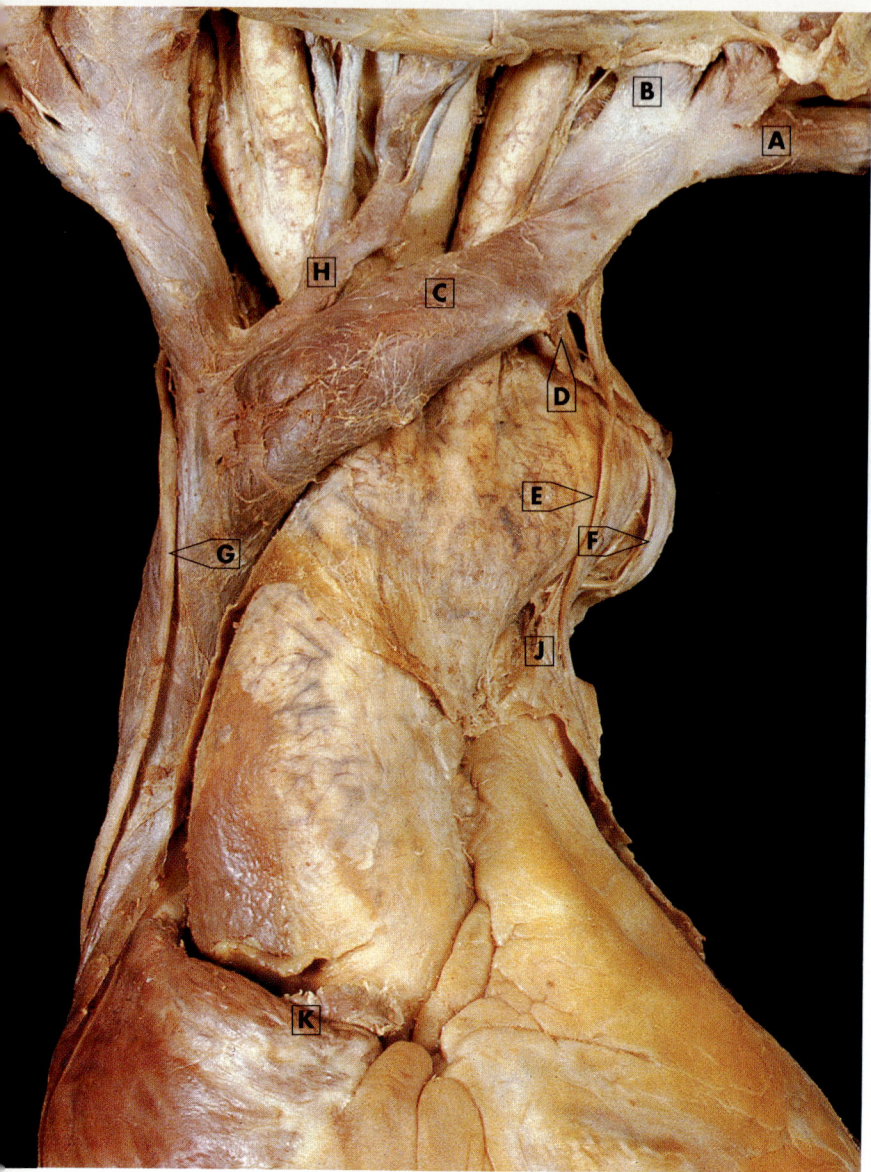

1. What terminates in the confluence of A and B?
2. What is the origin of D?
3. What is the inferior attachment of J?
4. What covers the surface of K?

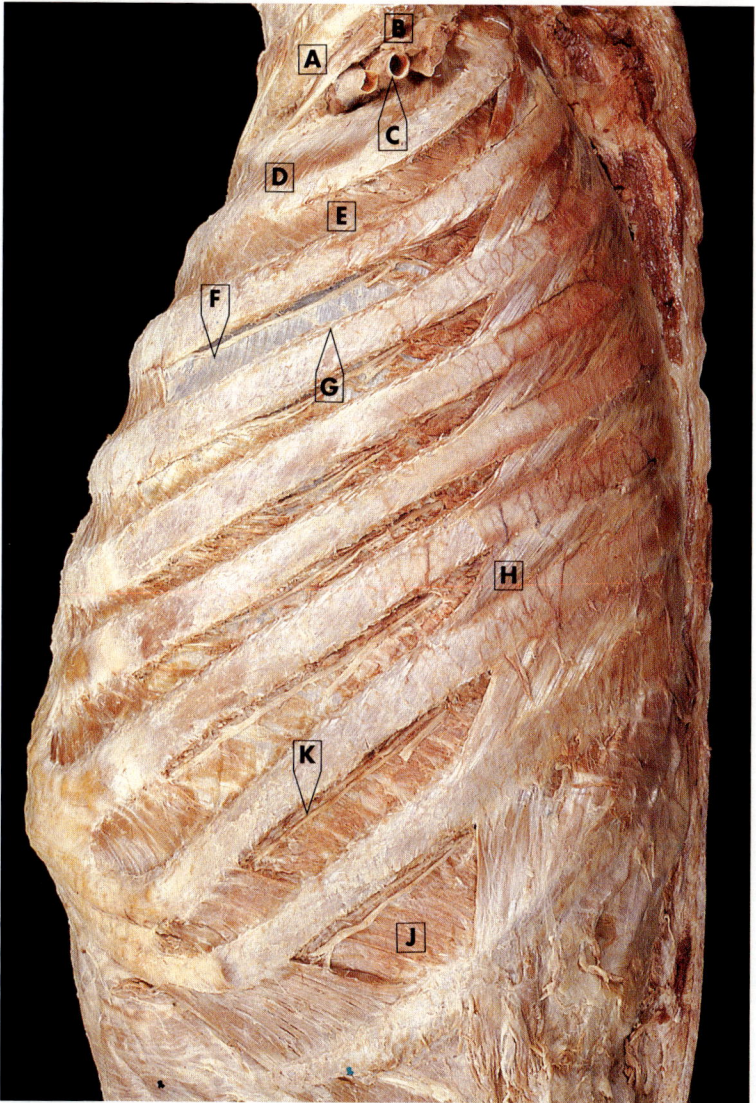

1. What is the action of D?
2. What are the actions of E?
3. What is the innervation of G?
4. What is the posterior origin of K?
5. What is the anterior origin of K?

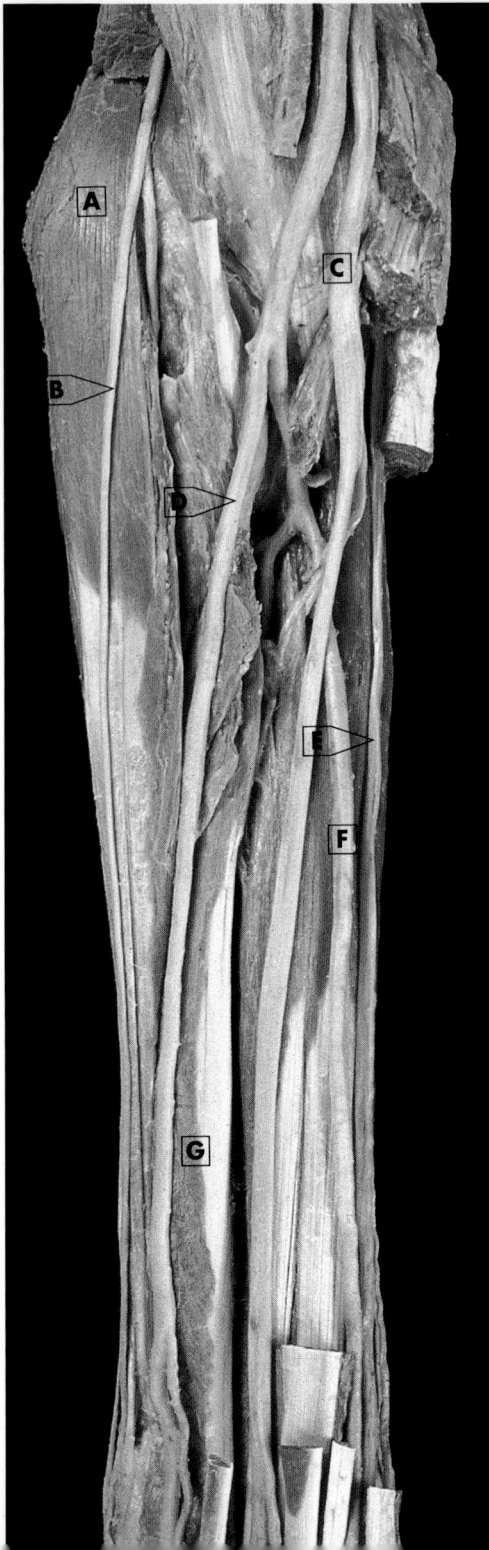

1. What does C supply in the arm?
2. How does D enter the hand?
3. How does E arise?

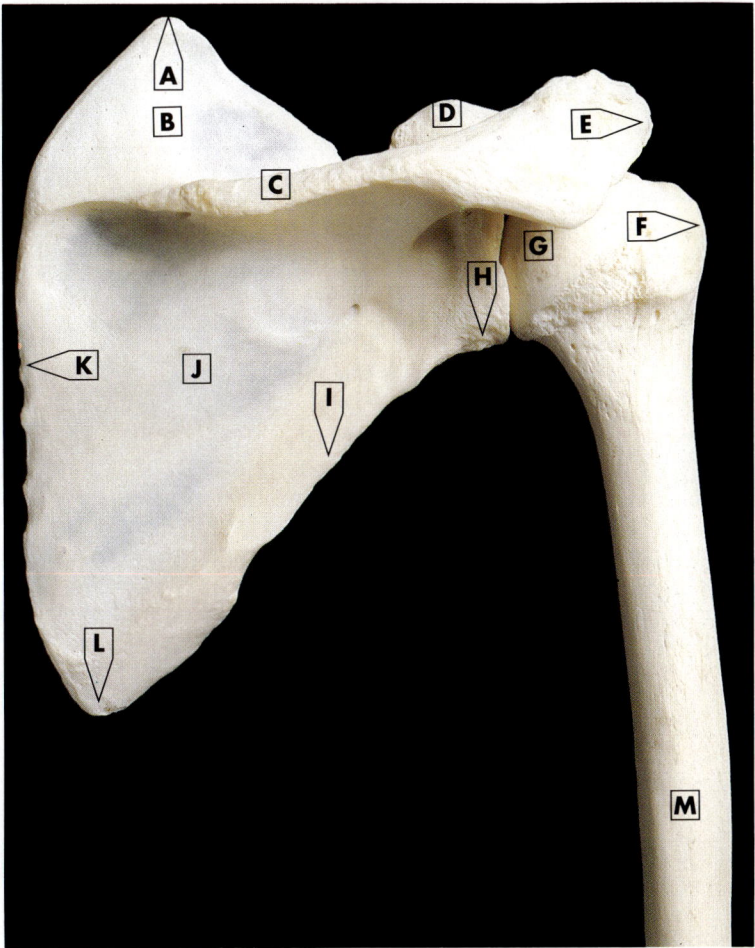

1. Classify the shoulder joint.
2. Name the rotator cuff muscles.
3. What attaches at H?

24

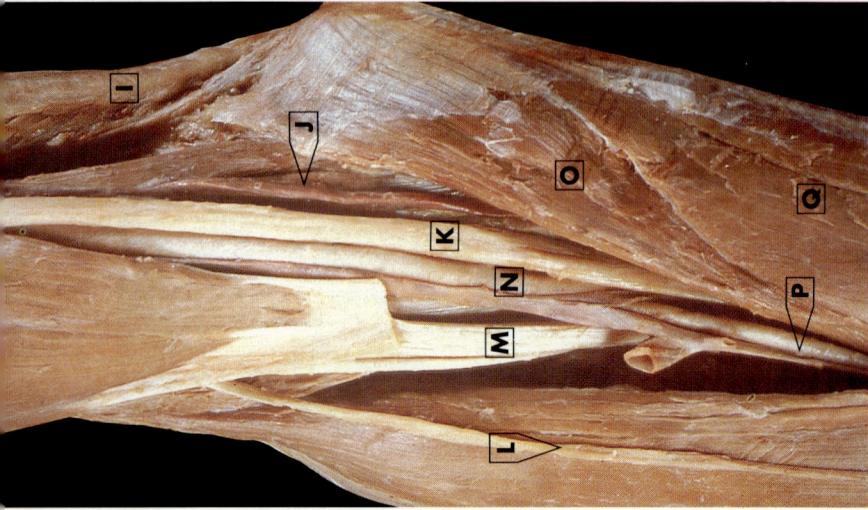

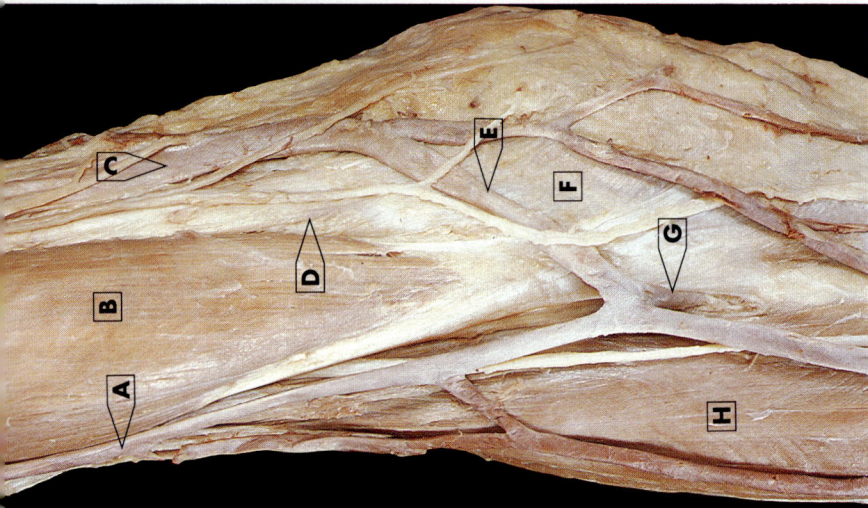

1. How does A terminate?
2. Name the space through which K & N are passing.
3. How does K arise?

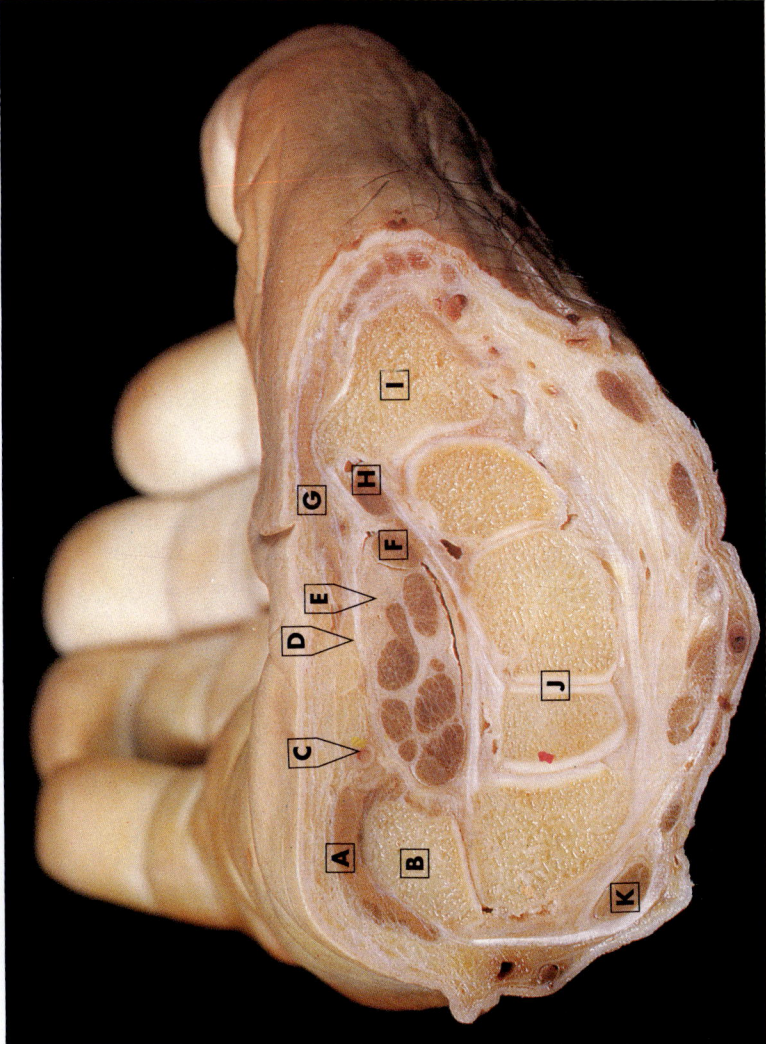

1. Which structure is most at risk from compression in "carpal tunnel syndrome"?

2. What is the nerve supply of A?

3. Which tissue is found at J?

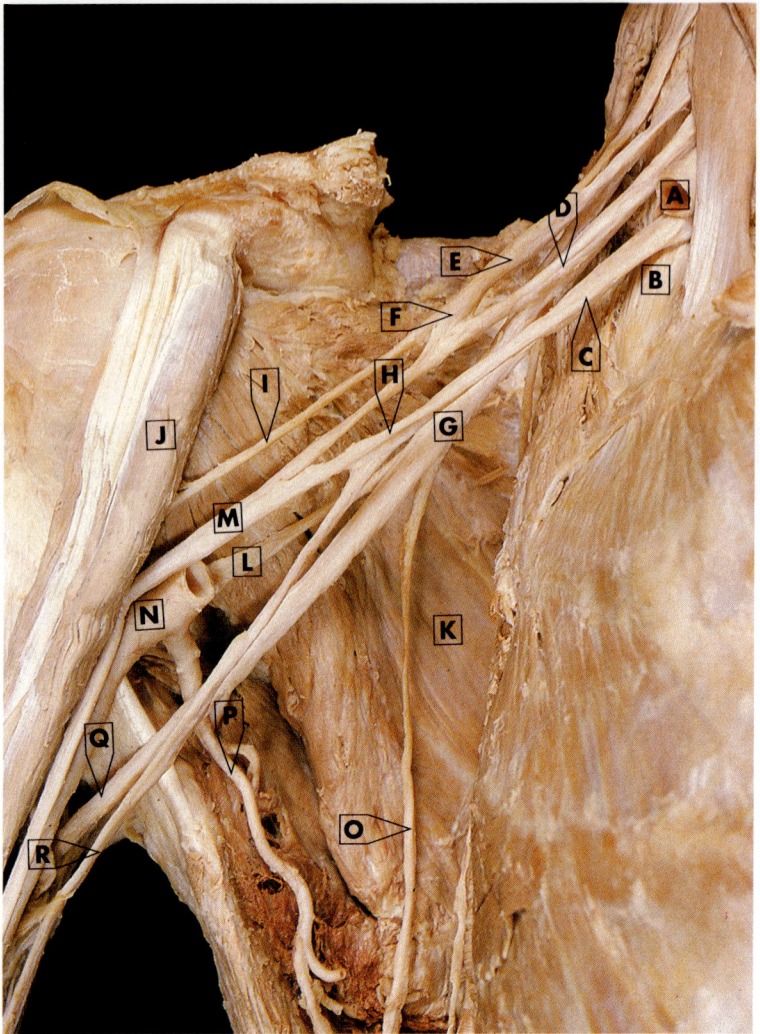

1. Which muscles form the anterior wall of the axilla?
2. What is the origin of A?
3. How does N terminate?

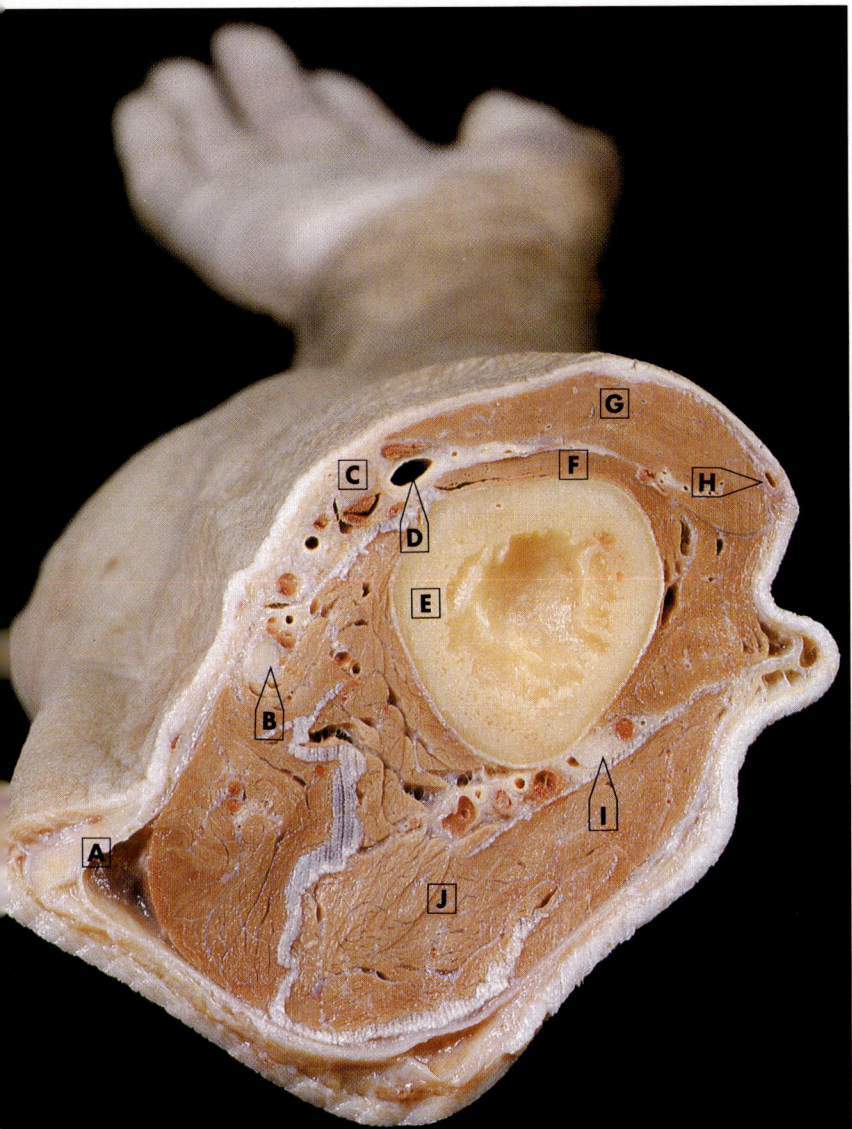

1. Which nerve supplies the muscles of the anterior compartment of the arm?
2. Give the origin of B.
3. What sign is characteristic of injury to I?

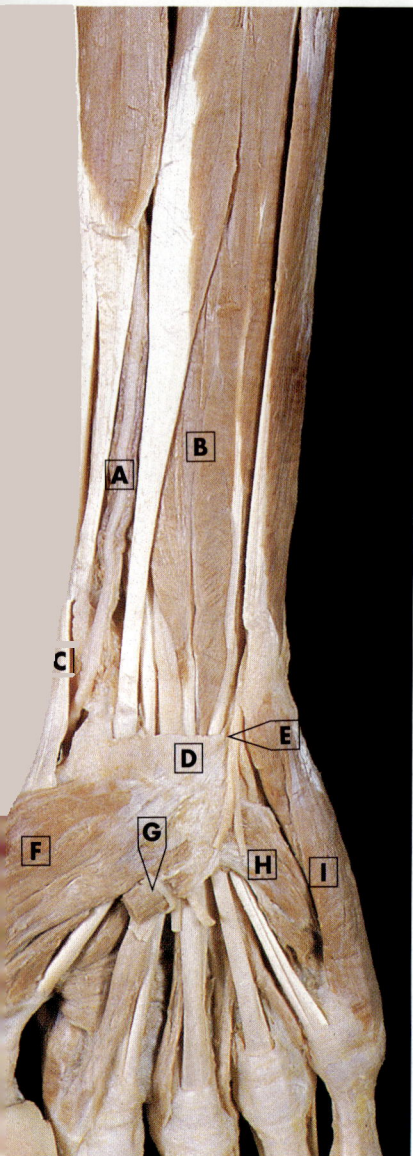

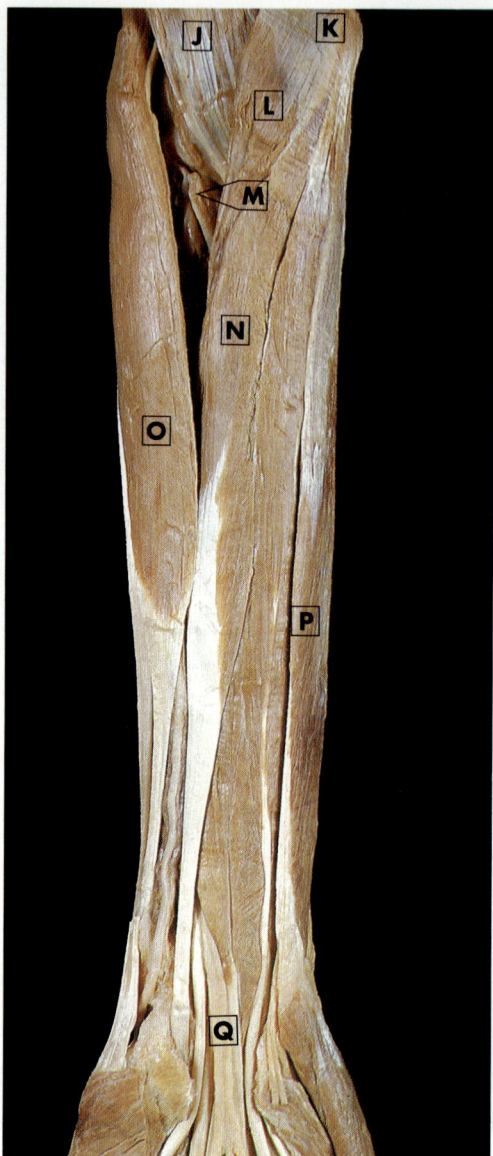

1. What covers the tendons deep to D?
2. Which nerve supplies O?
3. What is the distal attachment of P?

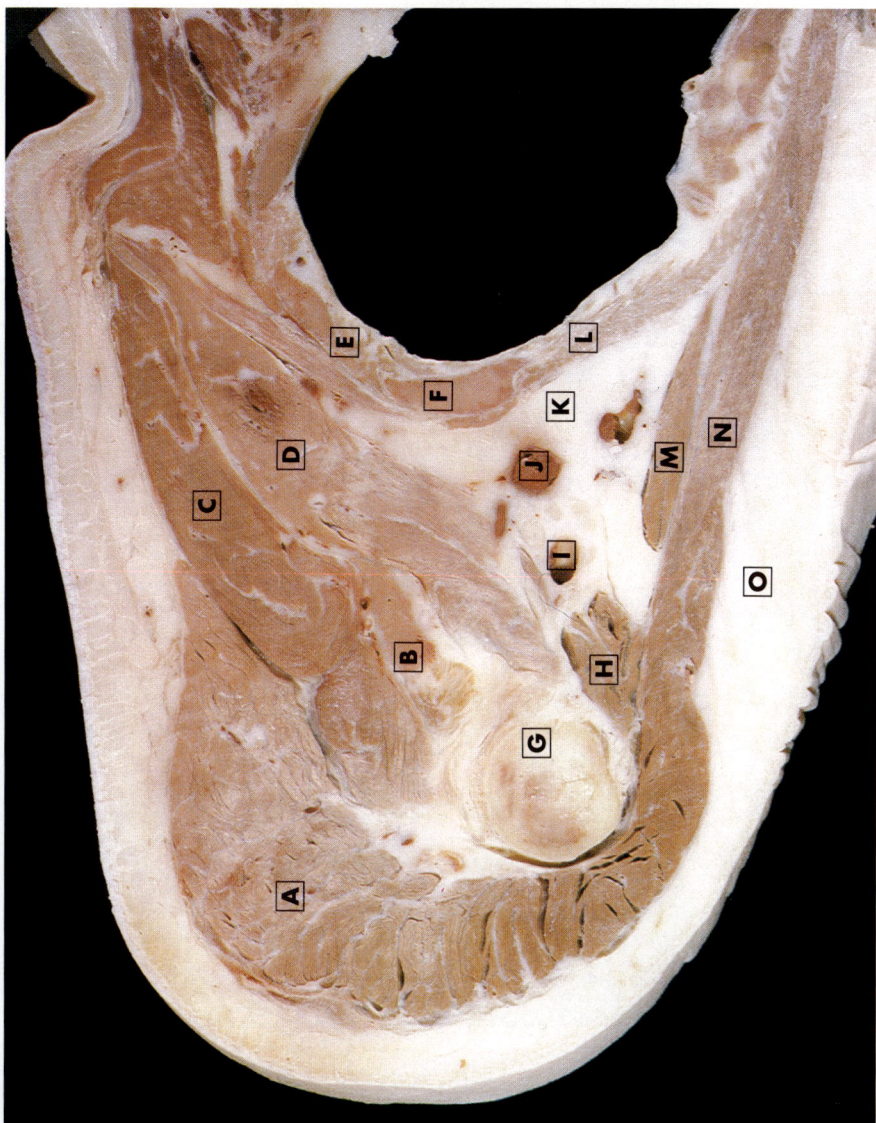

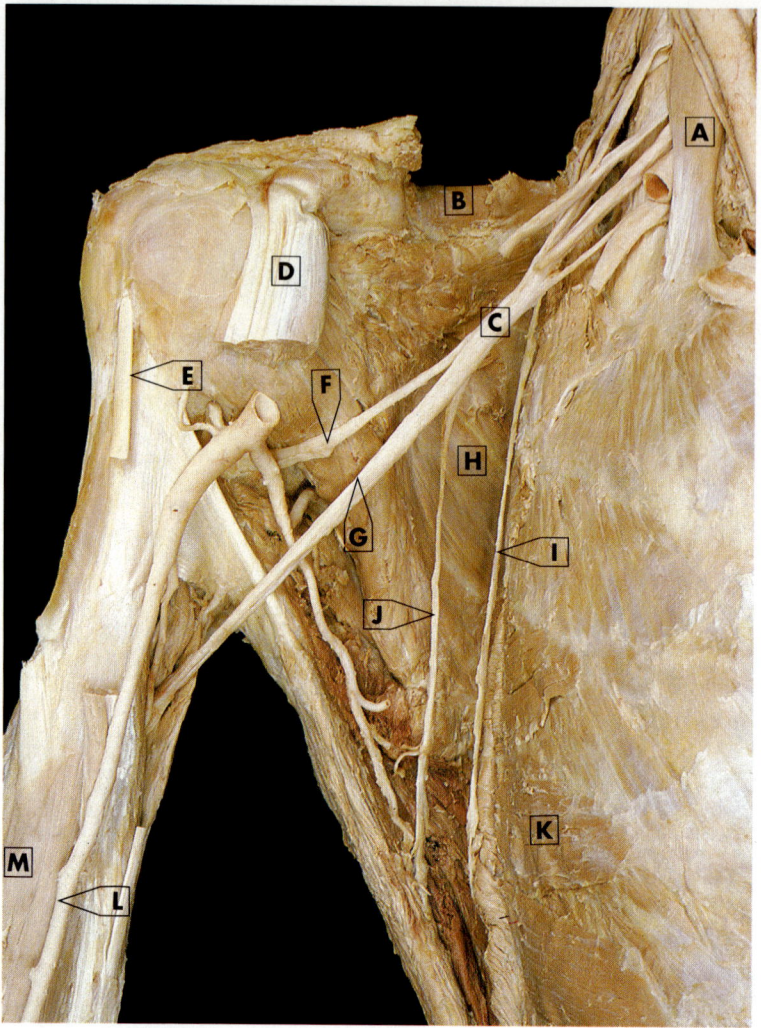

1. Give the origin of C.
2. What is the distribution of F?
3. What is the action of K?

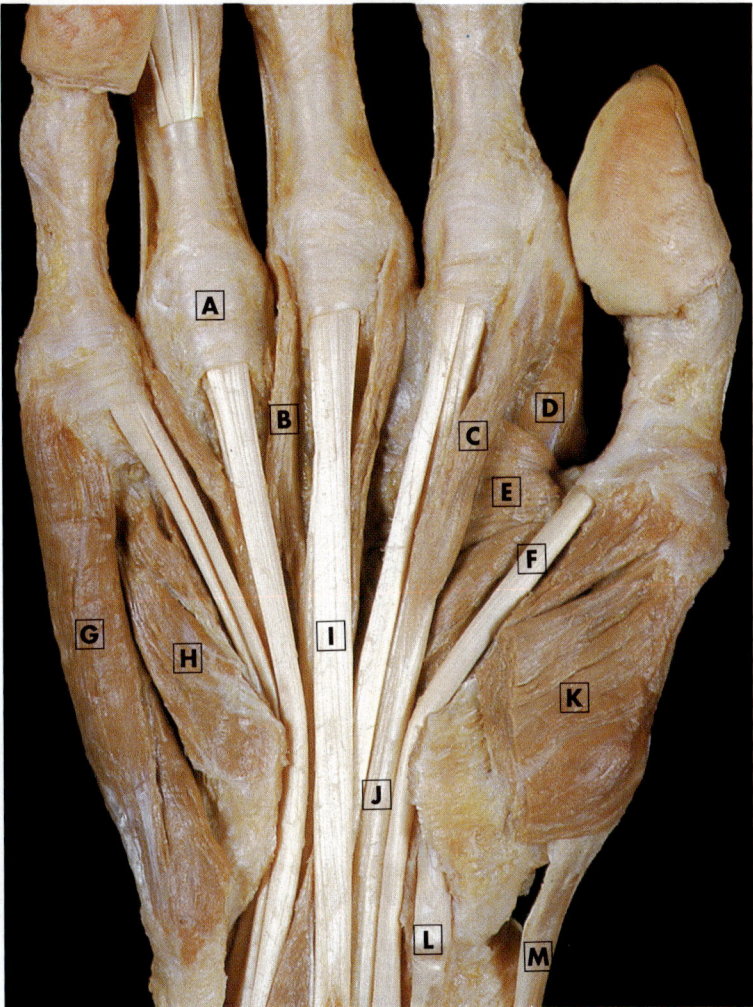

1. What is the nerve supply of D?
2. Which nerve supplies E?
3. How does I terminate distally?

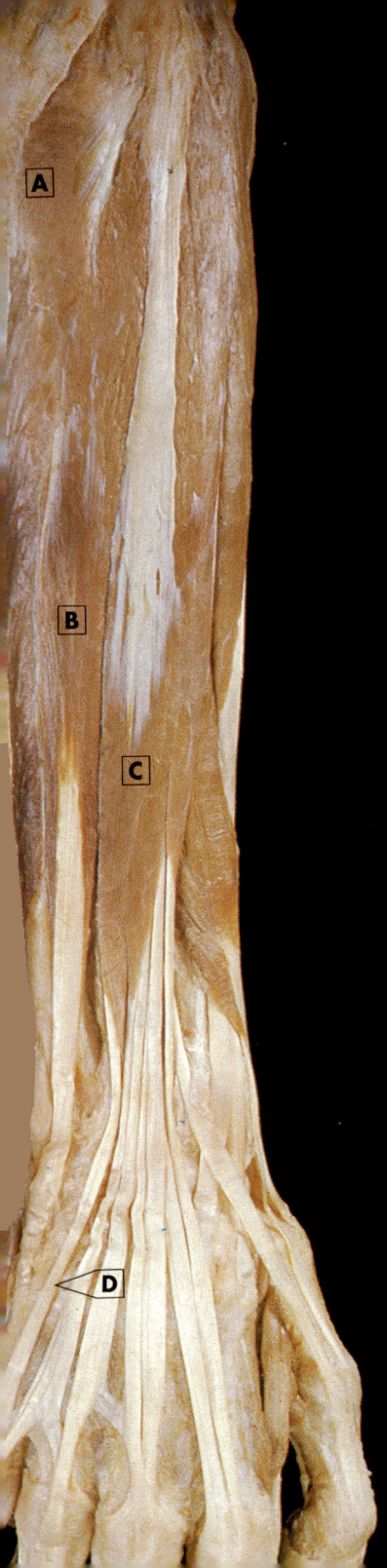

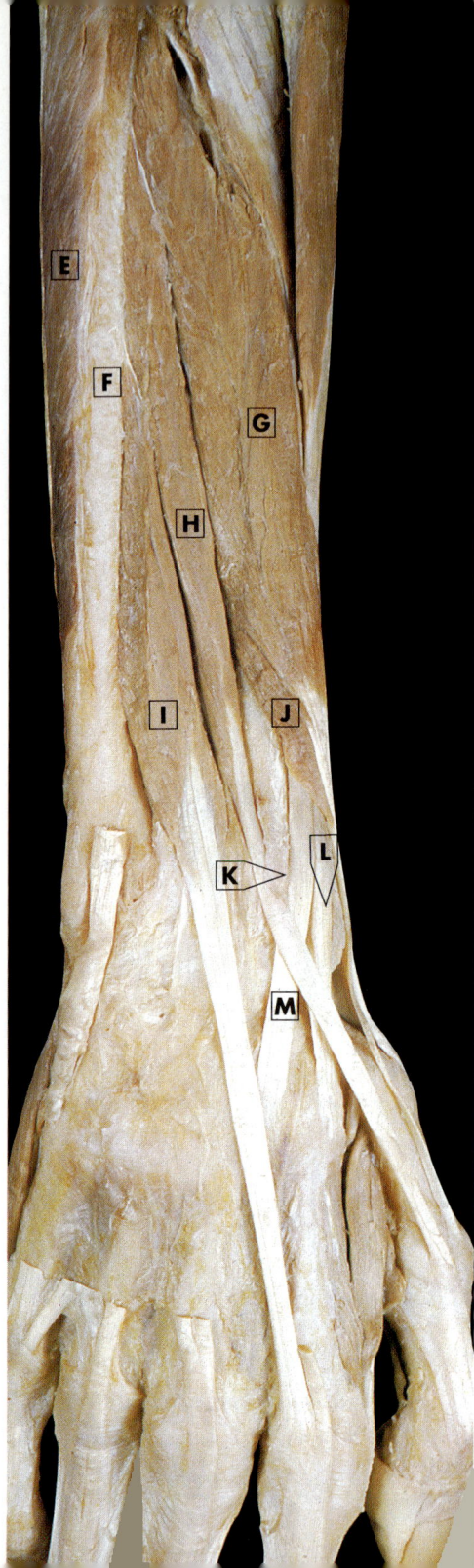

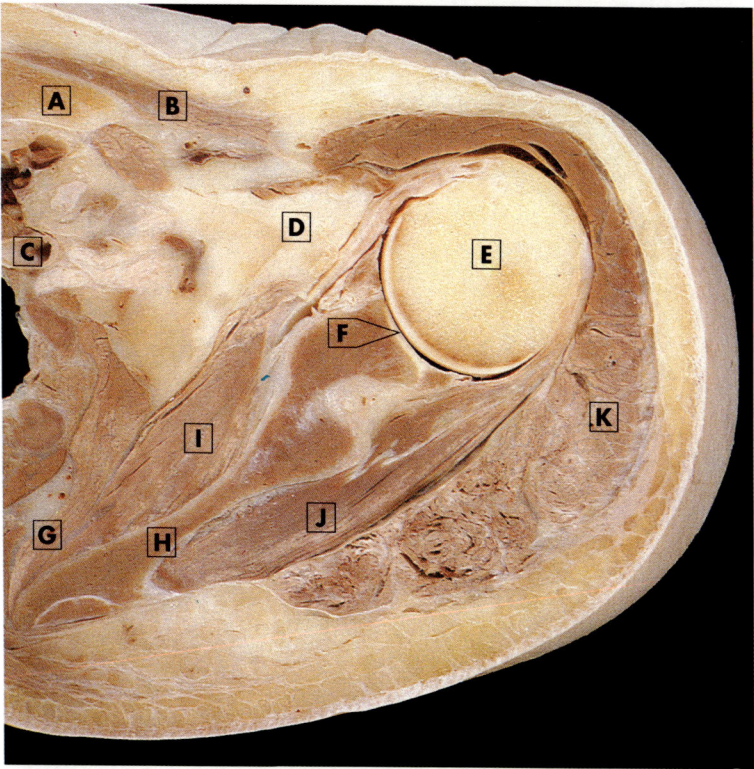

1. What is the nerve supply of K?
2. Which are the principal flexors of the shoulder joint?
3. Which bursa often communicates with the shoulder joint?

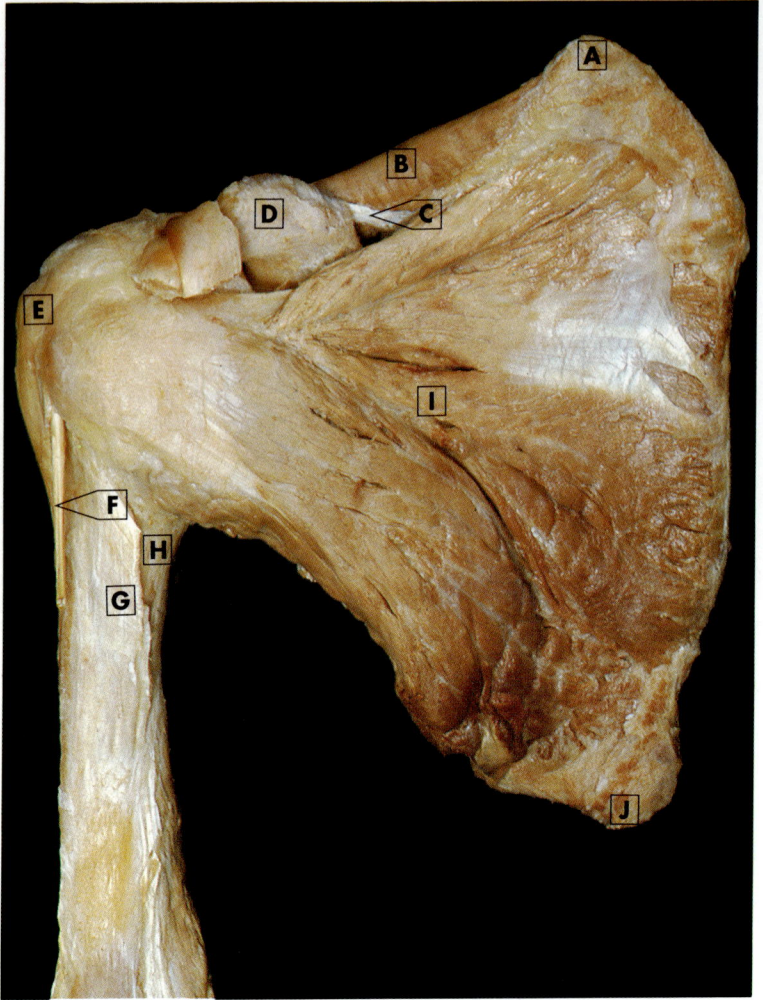

1. What is the action of the muscle attaching at G?
2. Which nerve is closely related at H?

3. What is the lateral attachment of I?

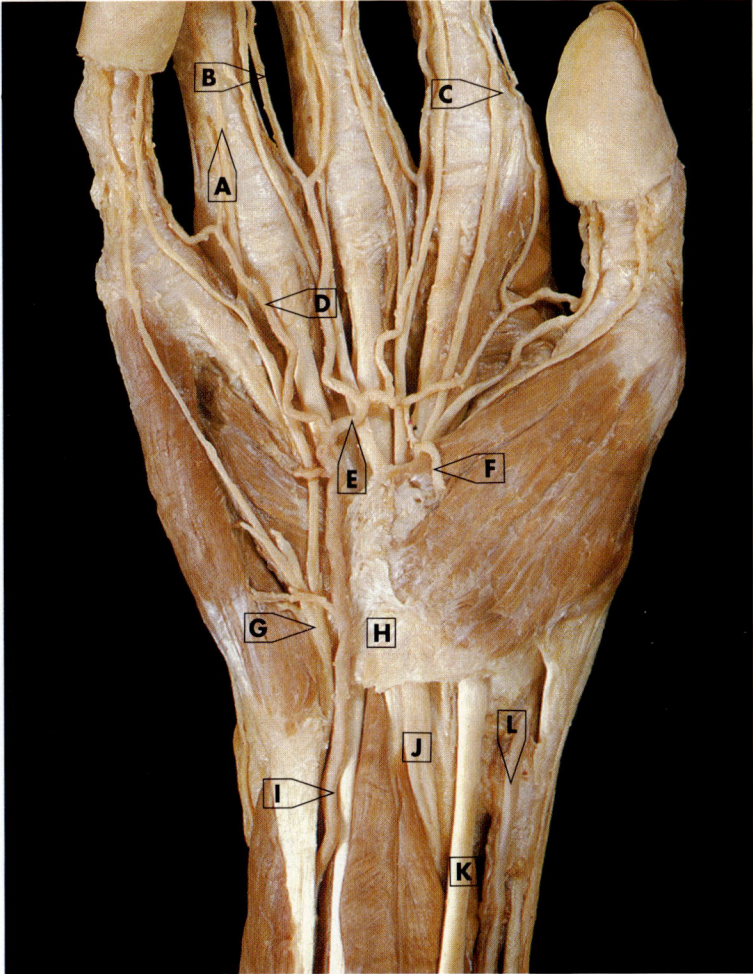

1. Which muscles are supplied by G in the forearm?
2. What is the origin of I?
3. How does L terminate?

35

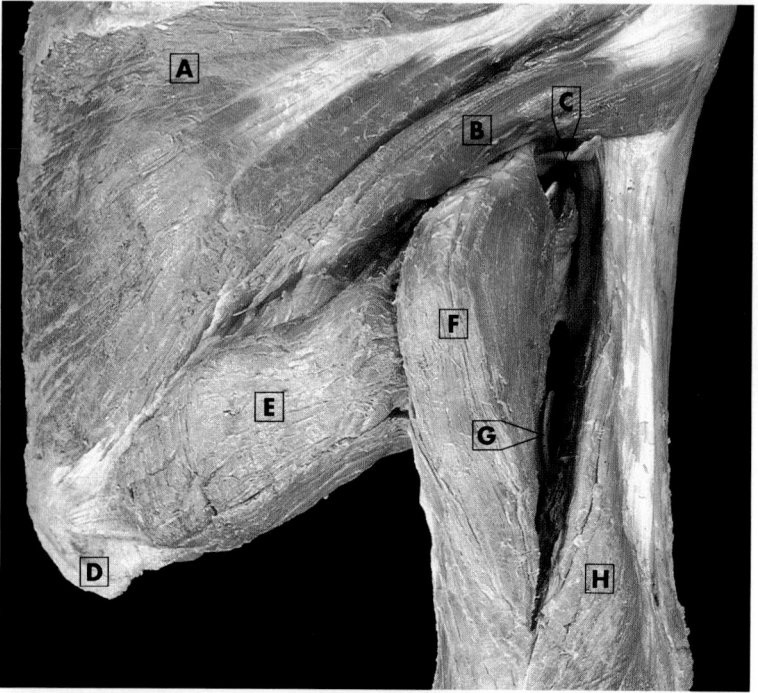

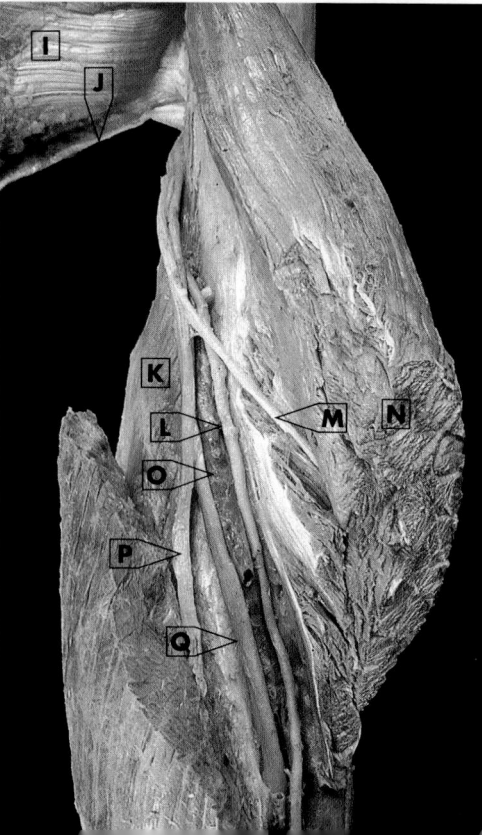

1. What is the nerve supply of A?

2. Where does E attach laterally?

3. What is the action of H?

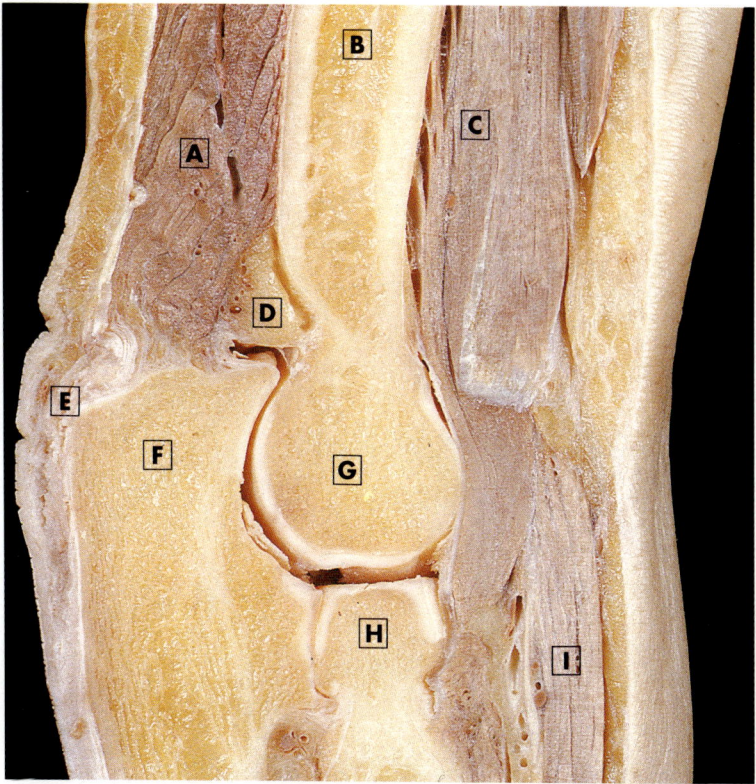

1. Which myotomes are associated with C?
2. Classify the joint between F & G.
3. Which ligament is closely associated with H?

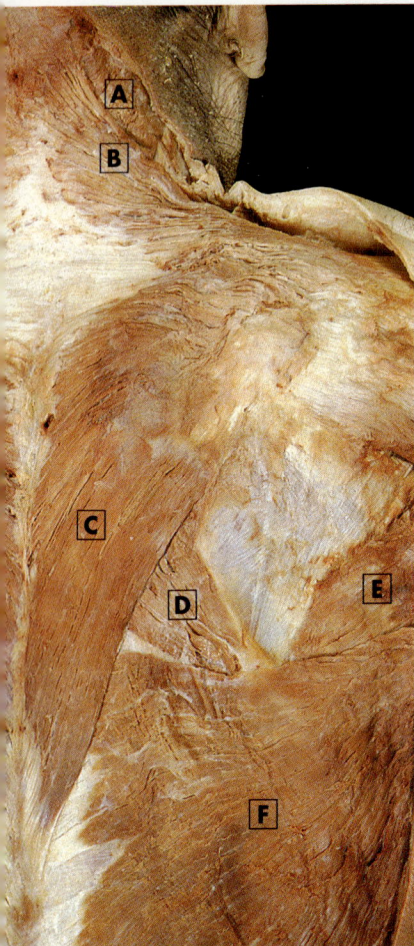

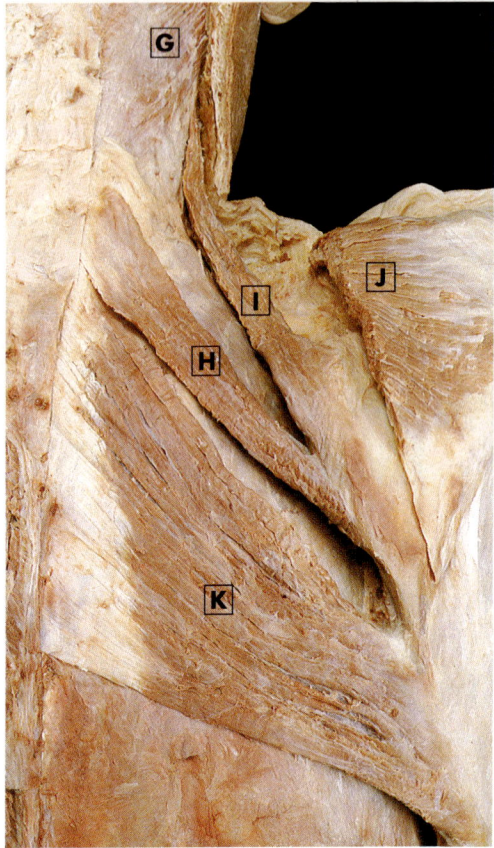

1. Give the lateral attachment of F.
2. What is the nerve supply of J?

3. What is the function of K?

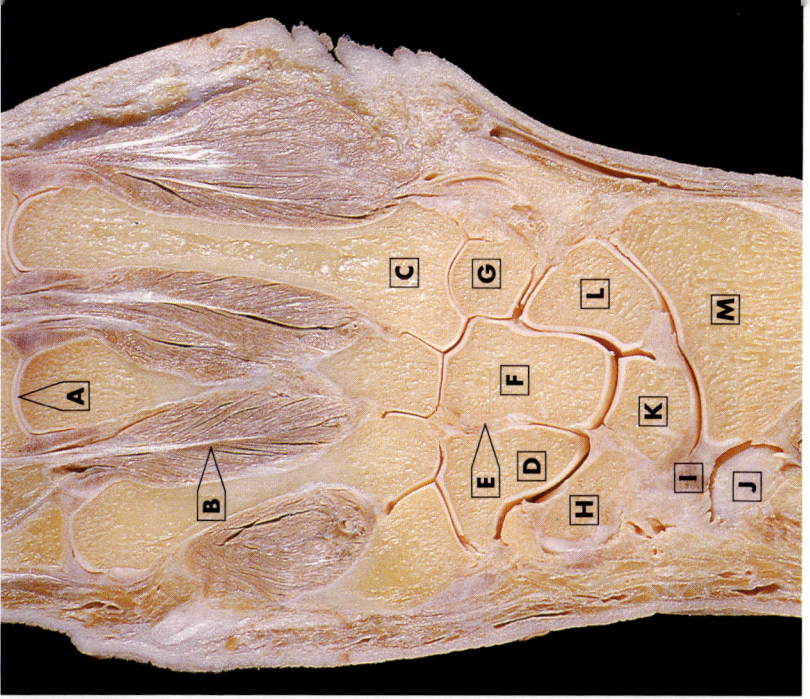

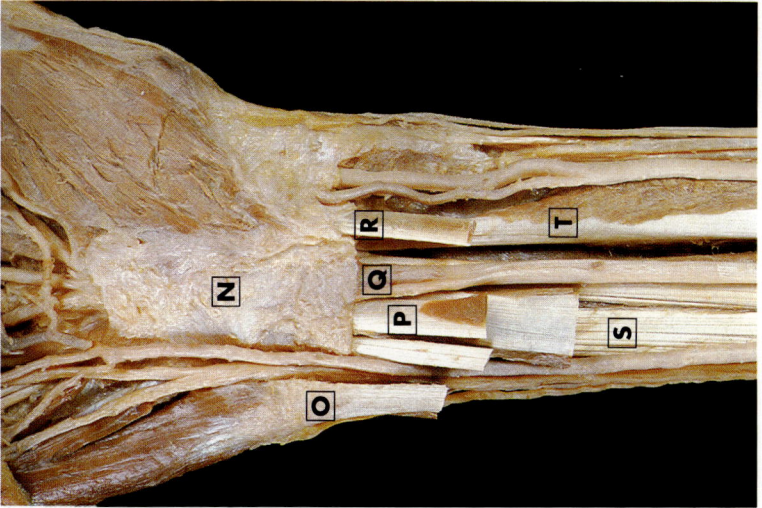

1. Give the nerve supply of B.

2. Where does R attach distally?

3. What is the proximal attachment of S?

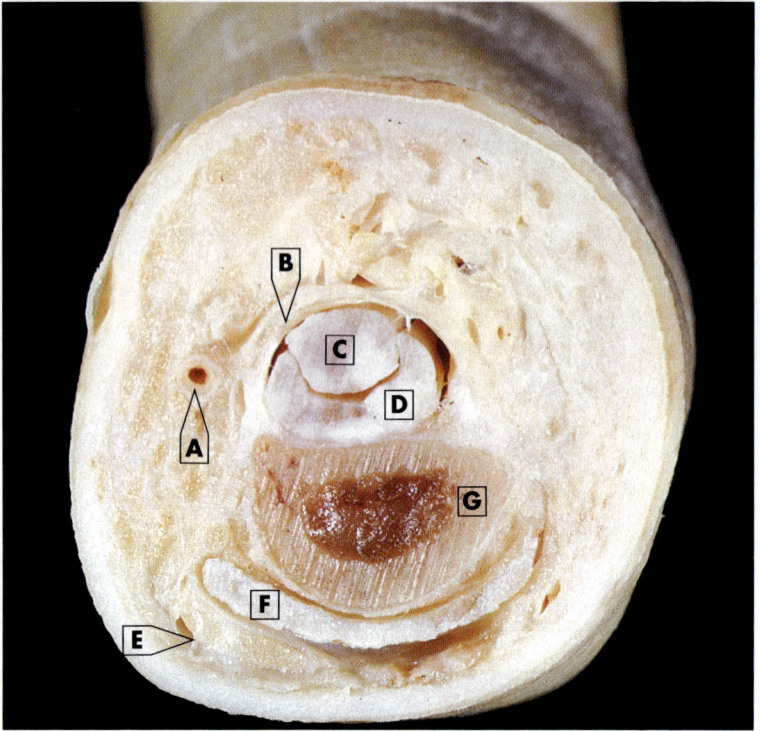

1. What is the level of this section?
2. How does E terminate?
40   3. Which intrinsic muscles attach to F?

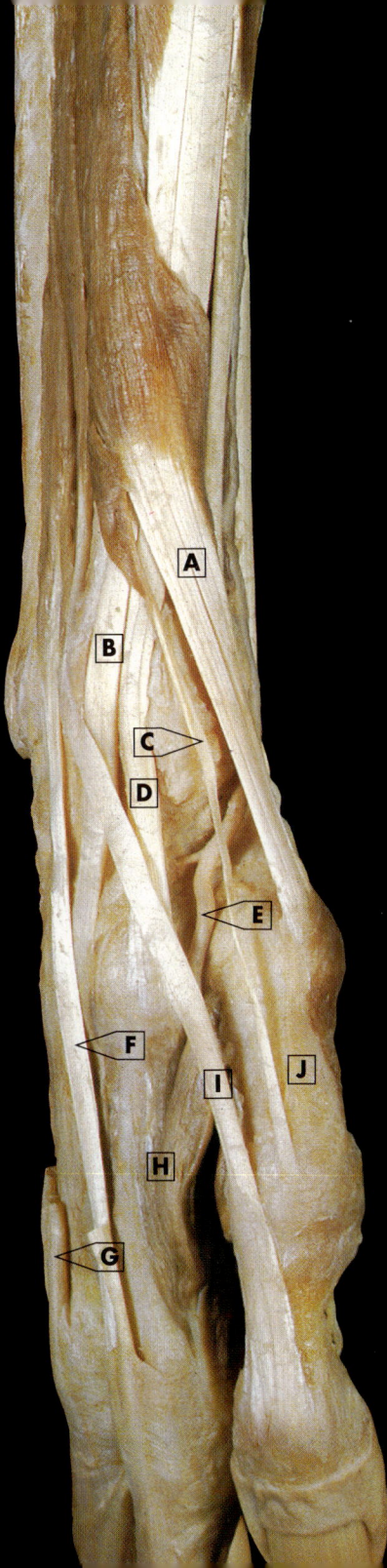

1. Name the interval between tendons A & I.
2. What crosses superficial to I?
3. What is the function of D?

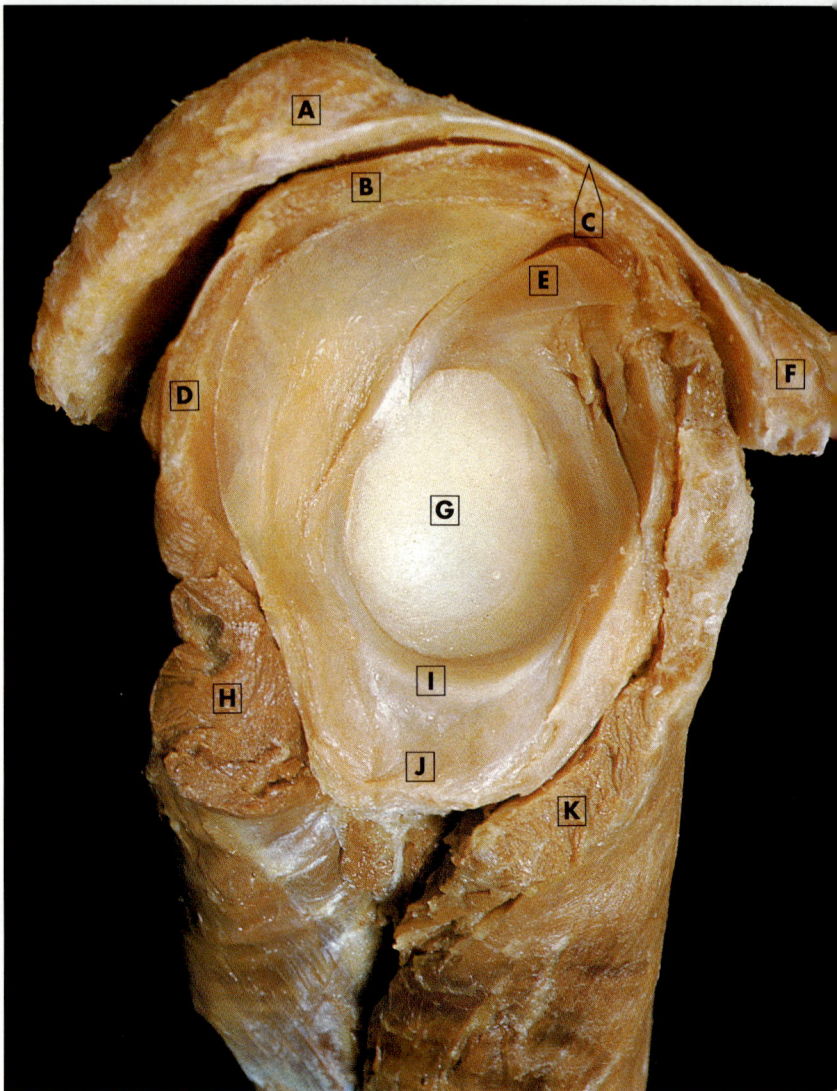

1. What lies immediately inferior to A?
2. What is the role of I?

3. Which tissue lines J?

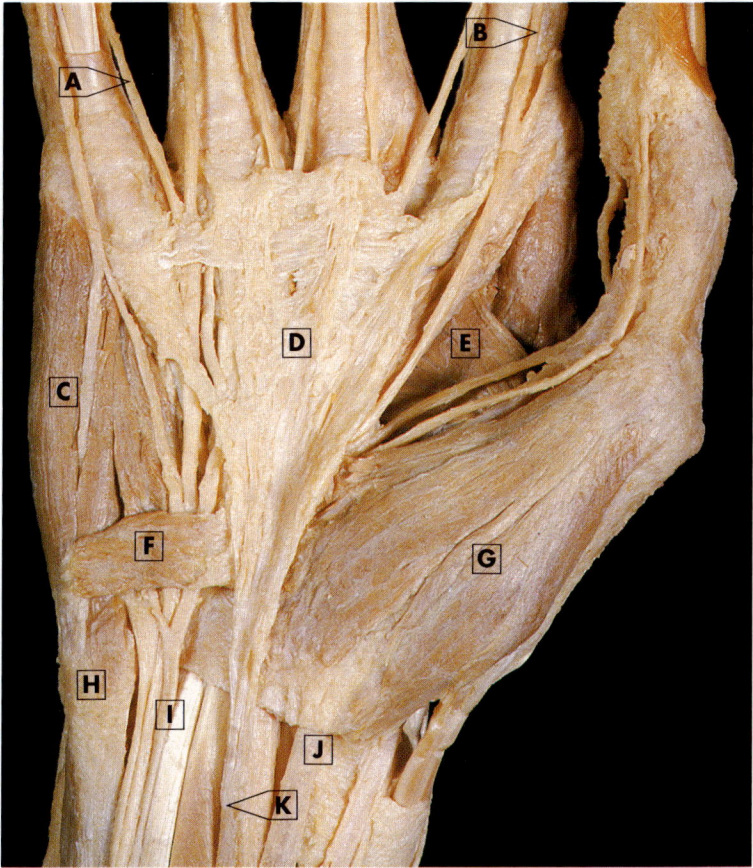

1. What is the function of D?
2. Give the origin of J?
3. Which dermatomes are associated with the palm of the hand?   43

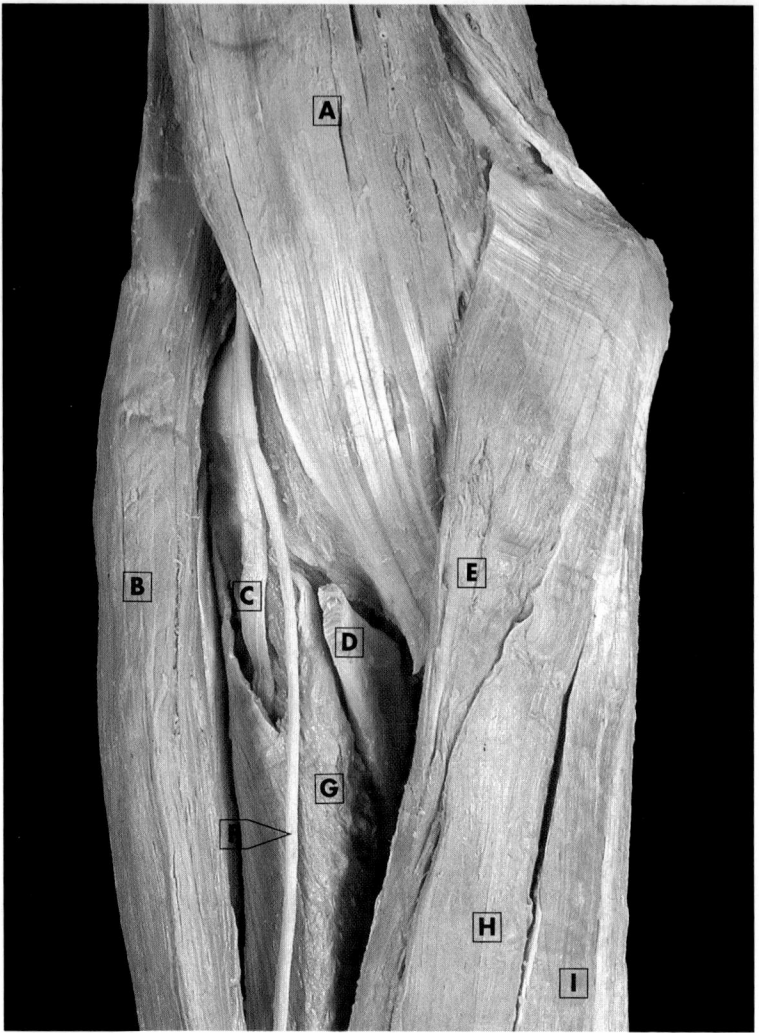

1. Which nerve supplies A?
2. What is the action of B?
3. Which muscles of the posterior compartment of the forearm are not supplied by C?

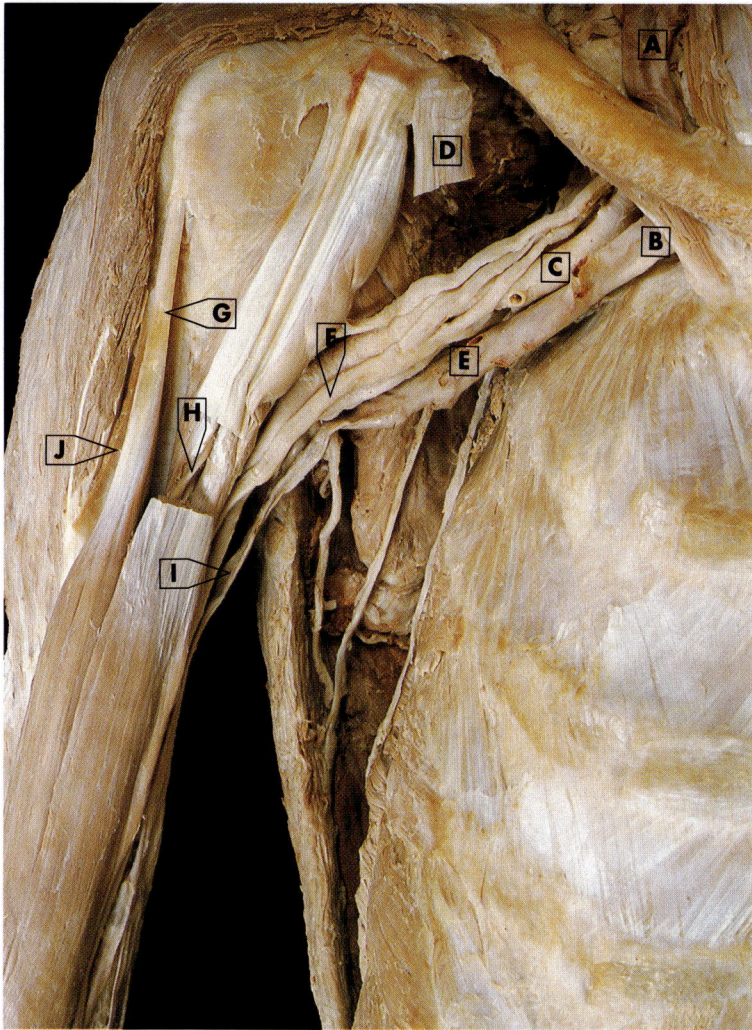

1. From which spinal cord segments does the brachial plexus arise?
2. Which structure invests C and the adjacent nerves?
3. What is the relationship of G to the shoulder joint?

45

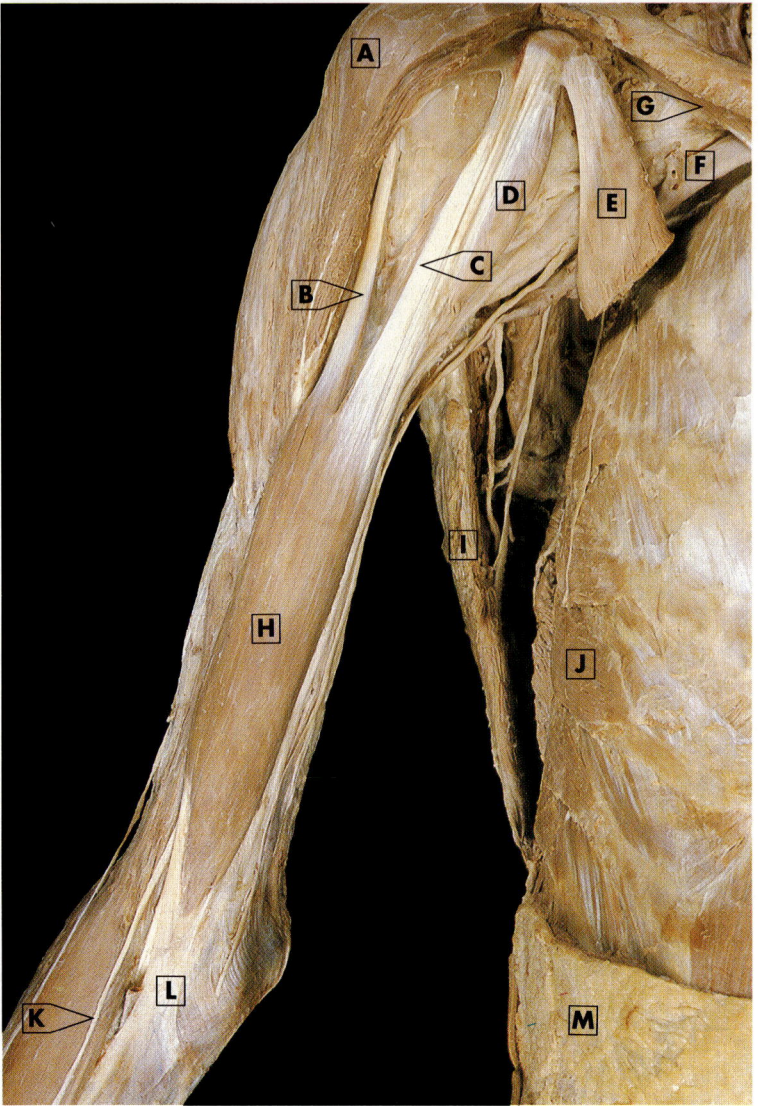

1. What are the main actions of H?
2. What is the origin of the nerve supplying J?
46 3. How does K arise?

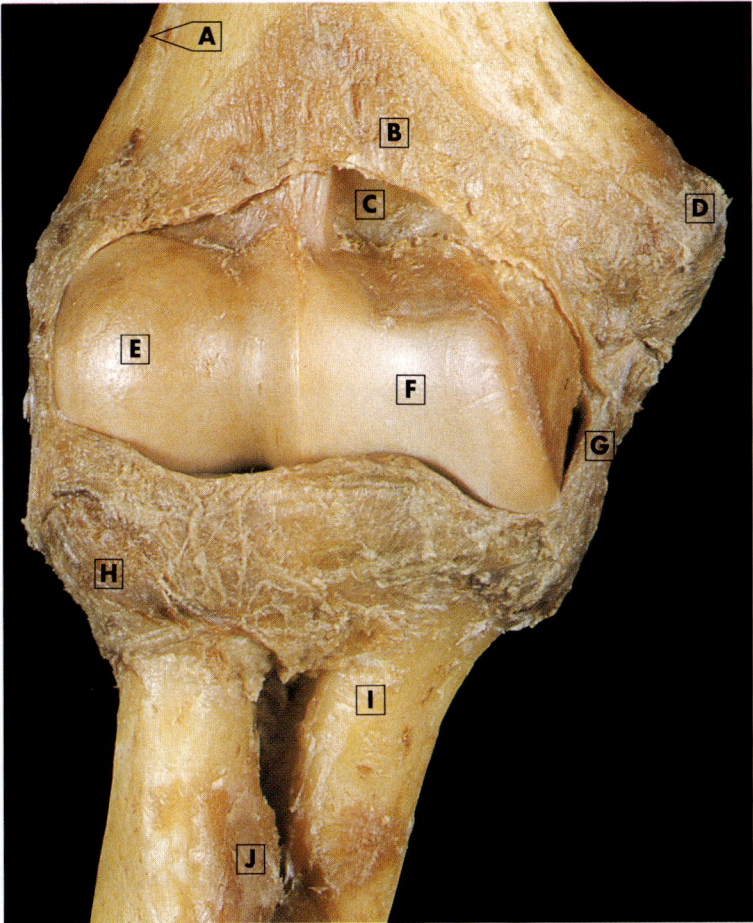

1. Which muscle attaches at A?
2. What attaches at I?
3. Which myotomes are associated with flexion of the elbow joint?

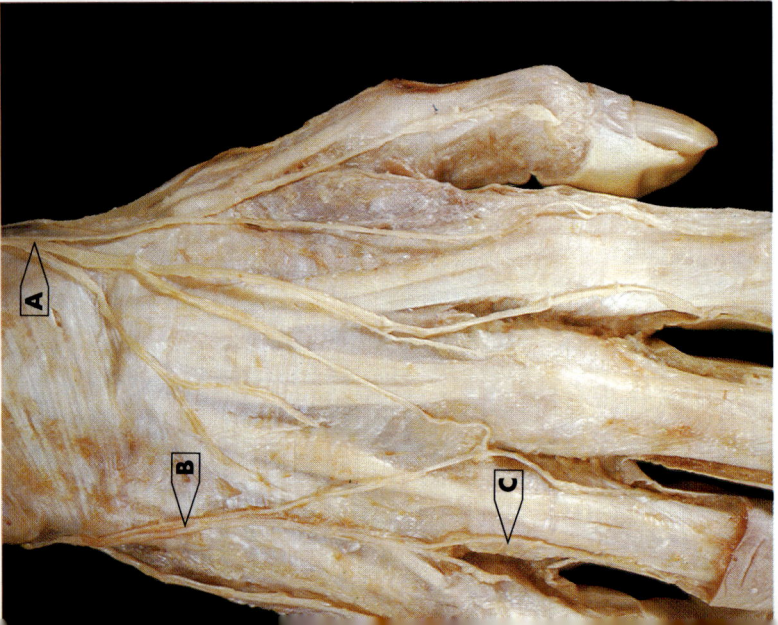

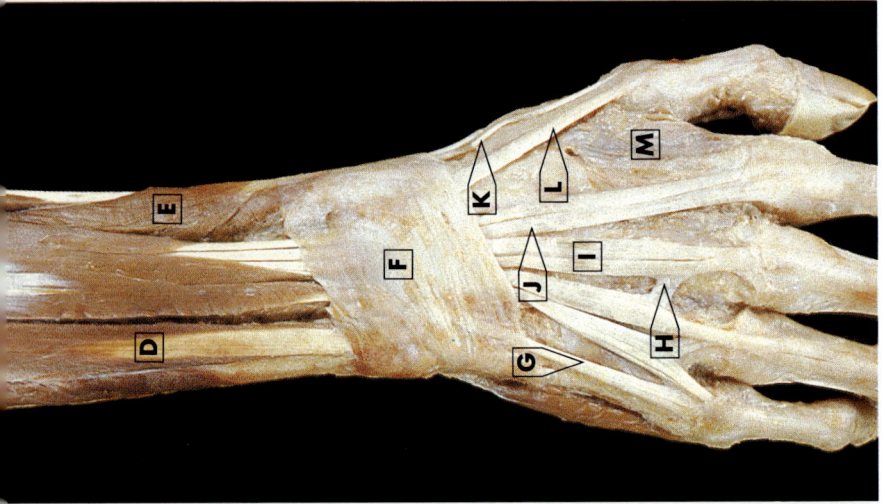

1. Where does D attach distally?

2. What nerve supplies E?

3. How does I terminate distally?

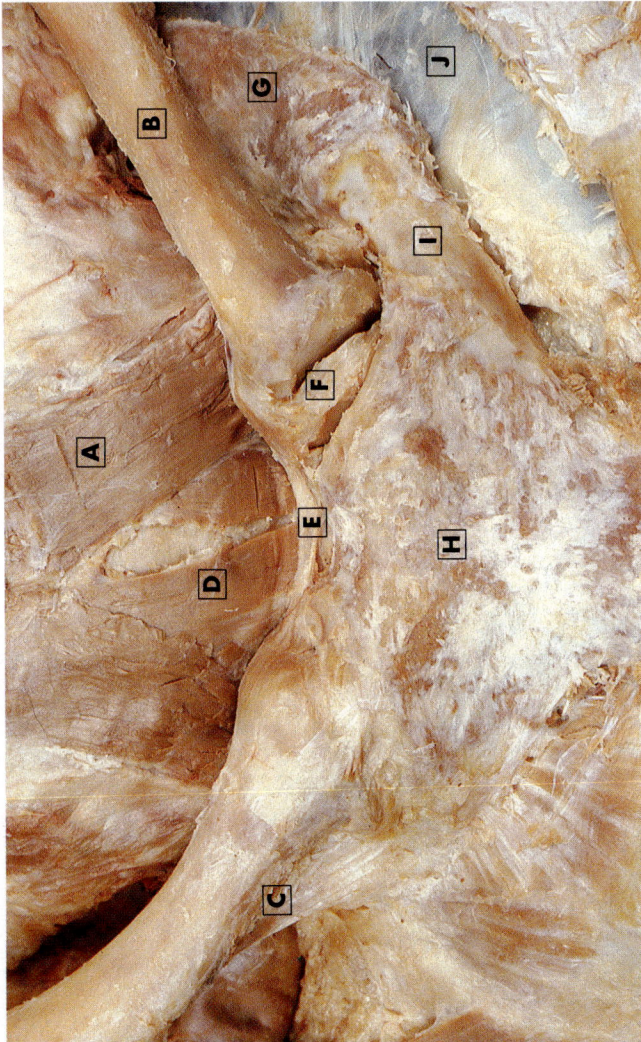

1. How does B ossify initially?

2. What is the role of F?

3. Which muscles have been removed to expose J?

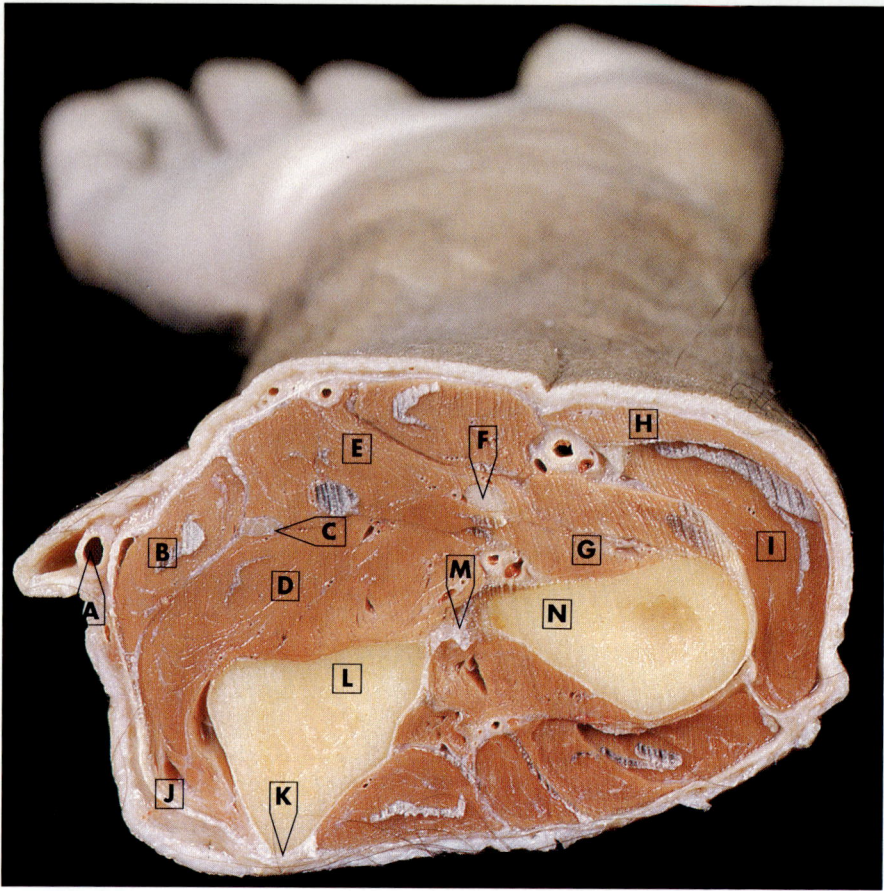

1. Which nerve(s) innervate(s) the muscles of the anterior compartment of the forearm?
2. What is the axis of movement between the bones of the forearm?
3. Which muscles does F supply in the hand?

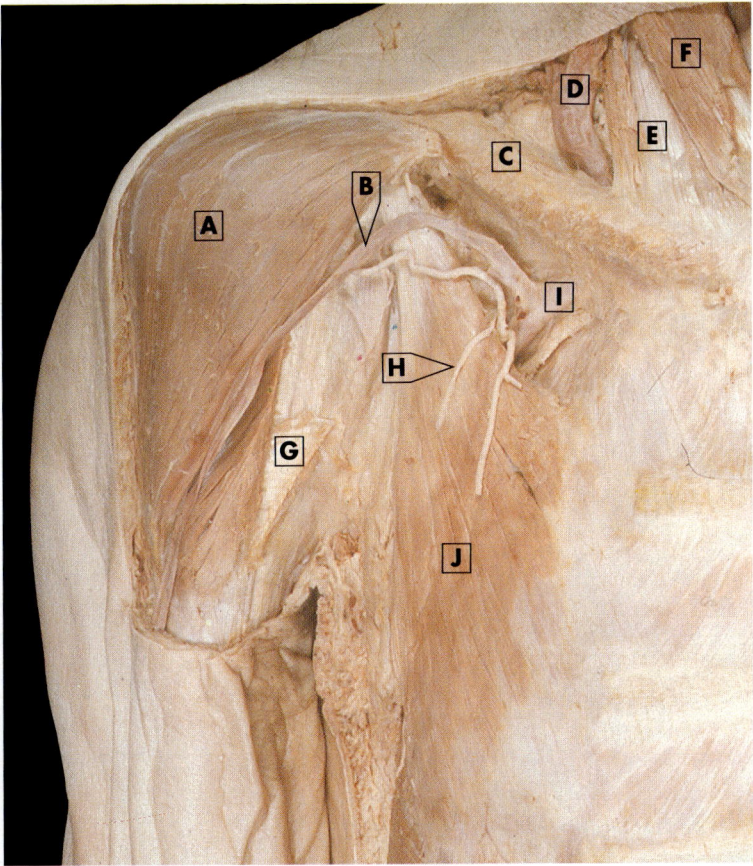

1. Give the distal attachment of A.
2. What is the origin of B?
3. What are the actions of J?

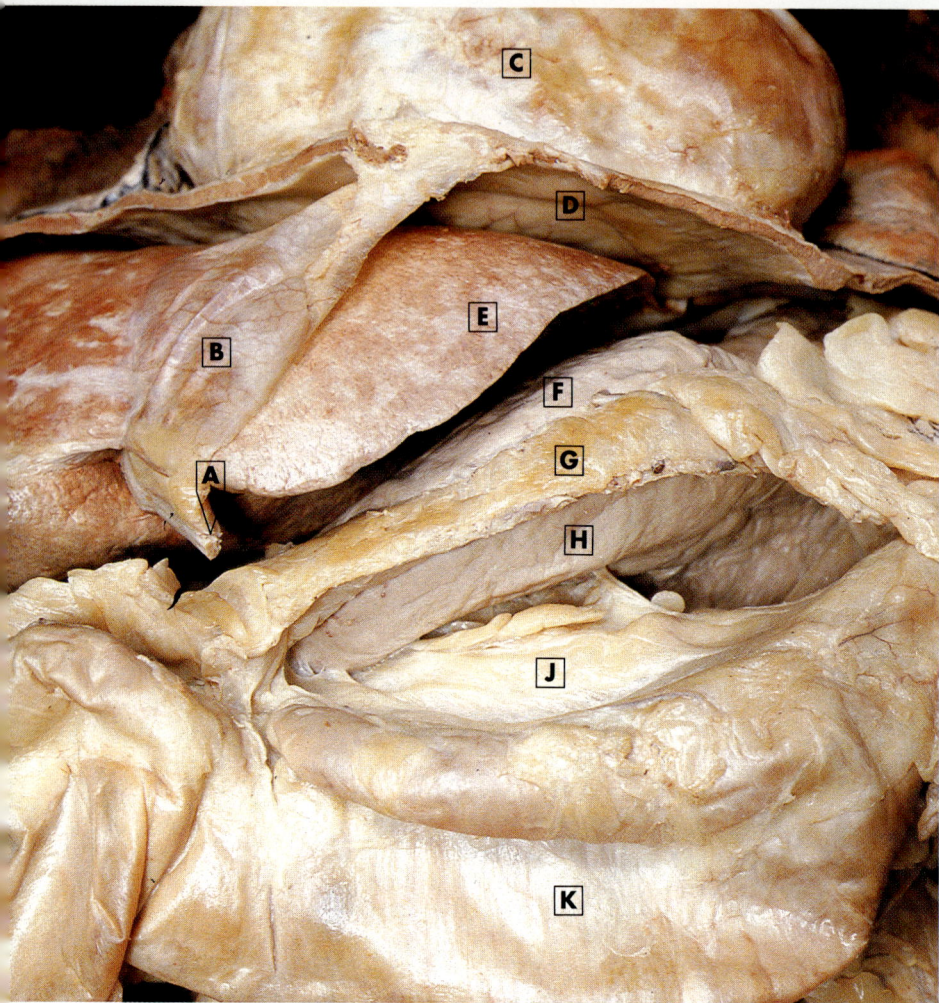

1. Of what embryological structure is A the remnant?

2. Name the potential space between D and E.

3. Name the potential space between H and J.

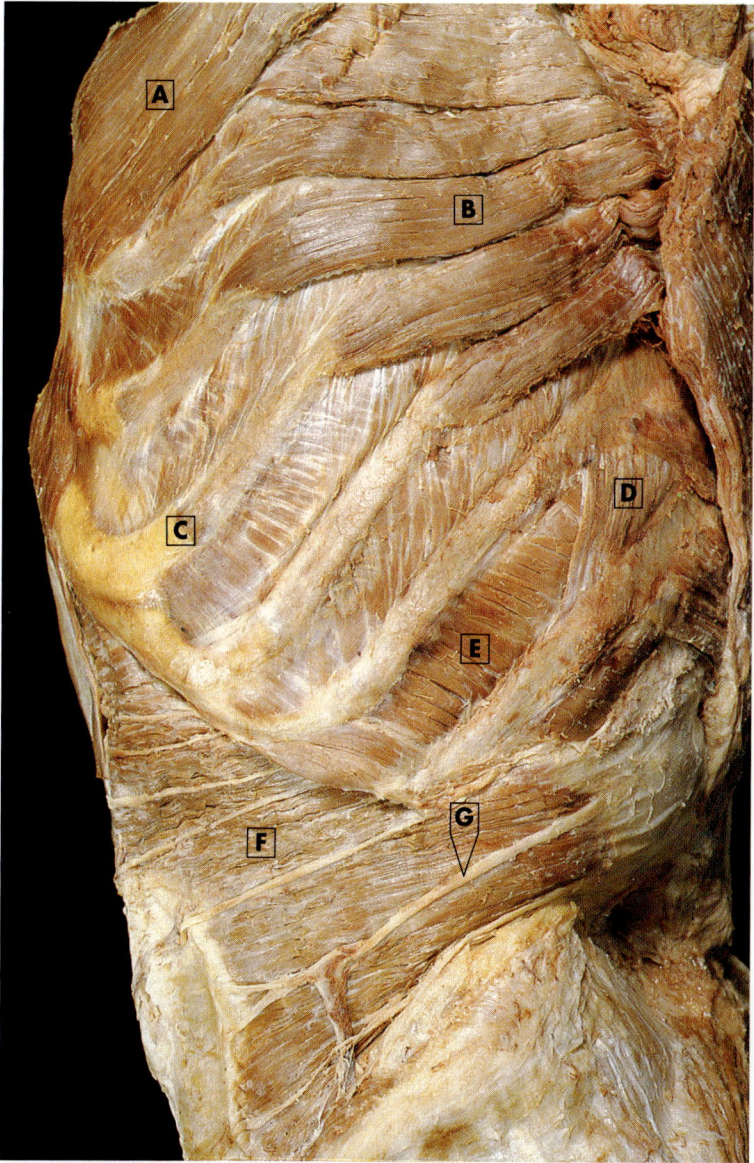

1. What are the actions of A?
2. What is the nerve supply to A?
3. What are the actions of B?
4. What is the nerve supply to B?
5. What is the segmental origin of G?

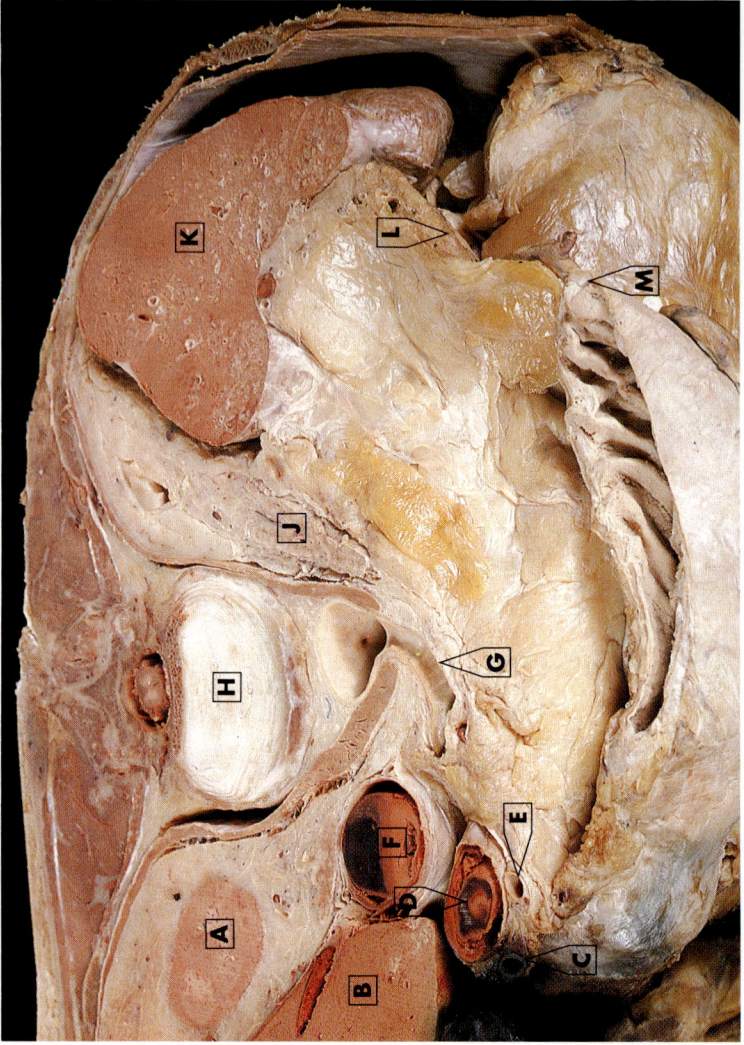

1. What is the termination of C?

2. What is the origin of D?

3. Name the space between D and F.

4. What type of joint is H?

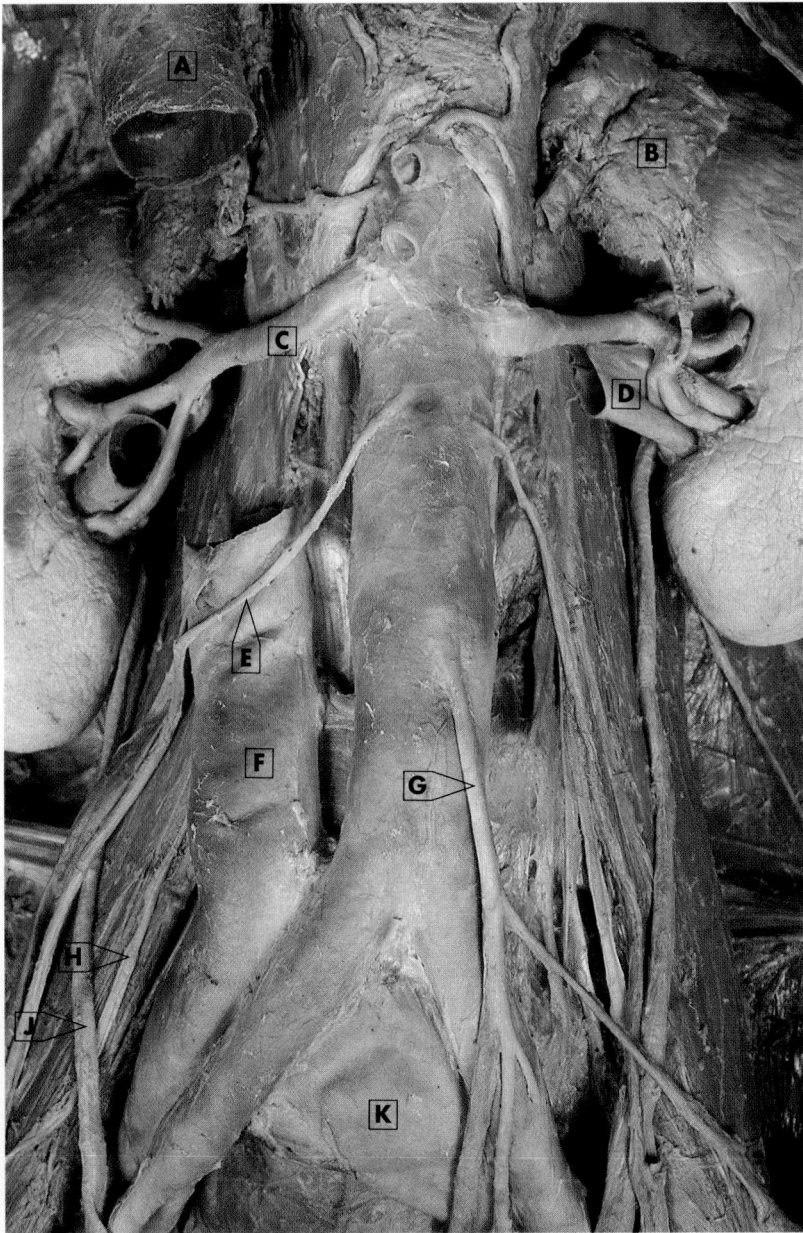

1. What are the anterior relations of B?
2. What organs are drained by D?
3. What is the distribution of H?
4. What is the distribution of G?

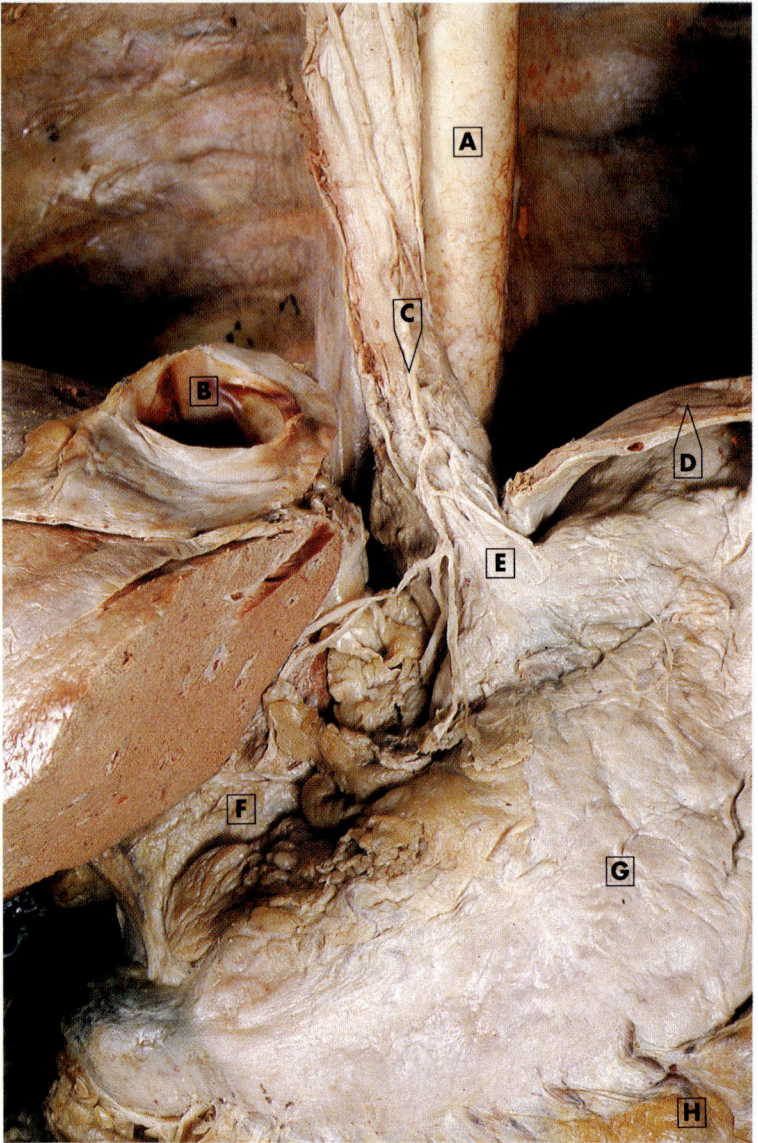

1. What branches does A have?
2. What is the origin of C?
3. What is the nerve supply to D?

4. What lies posterior to F?

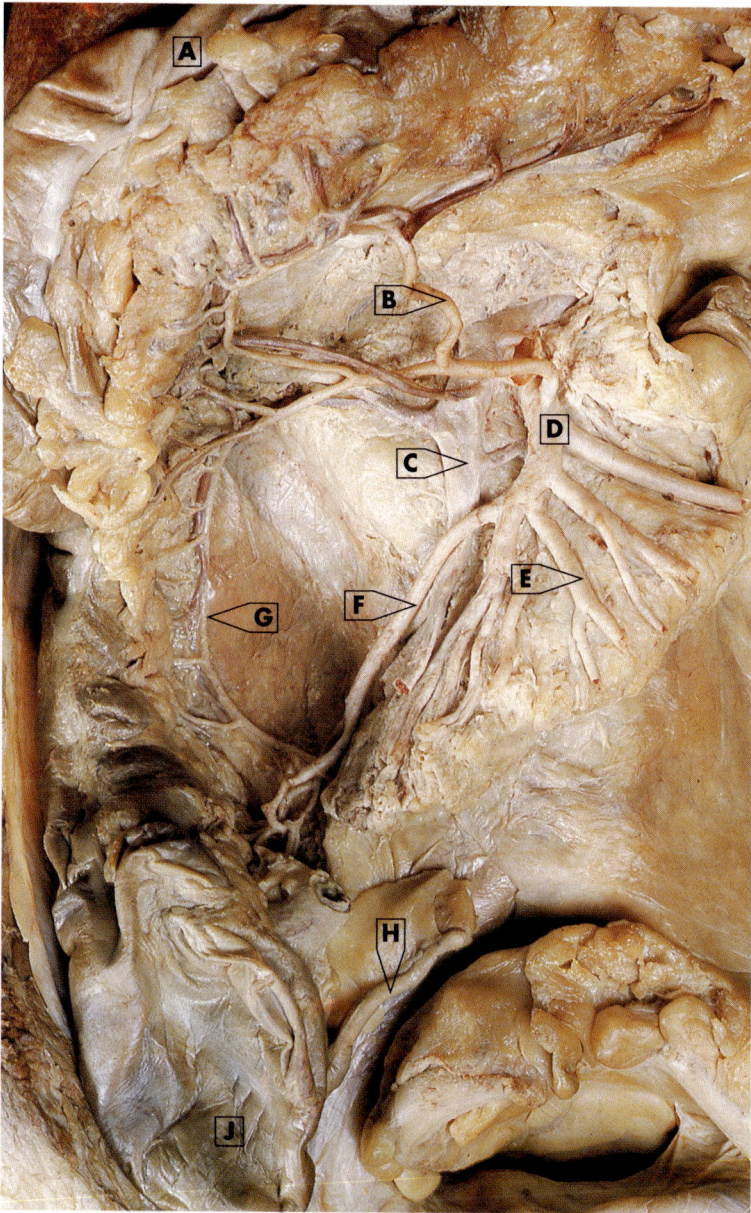

1. What is the termination of C?
2. What lies posterior to J?
3. The tip of H may lie in contact with which organs?

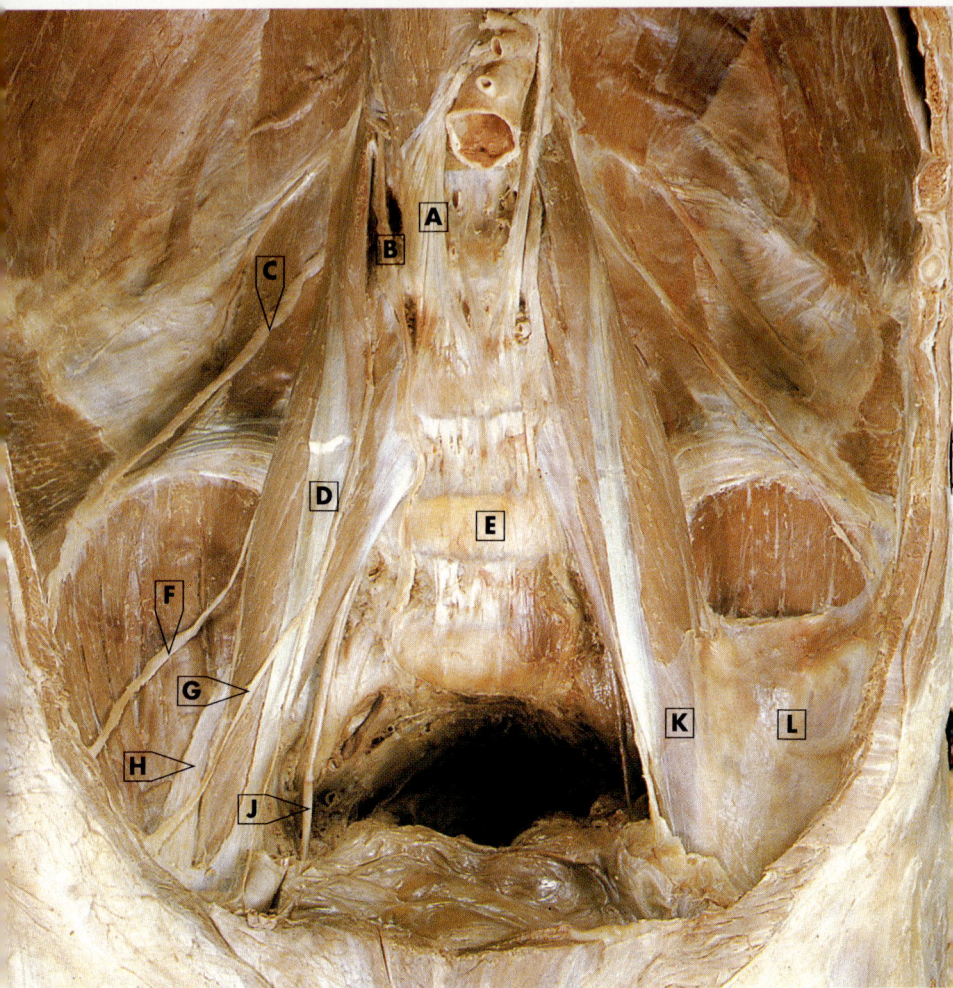

1. What is the termination of B?
2. What is the segmental origin of F?
3. Name the muscles innervated by H.
4. What is the distribution of J?

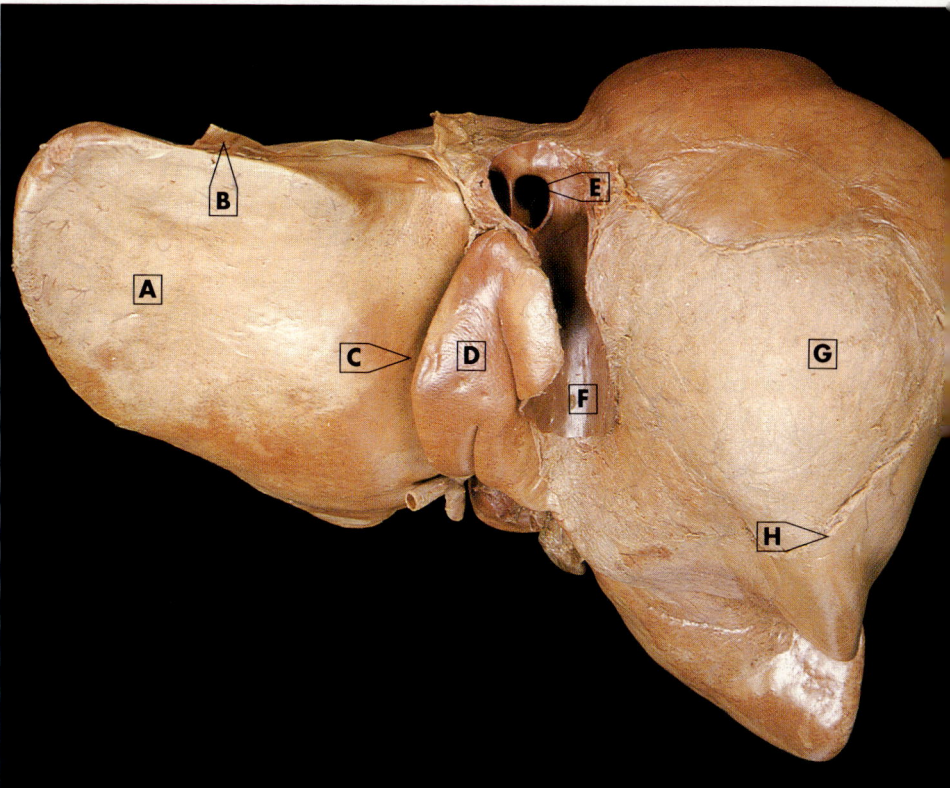

1. What is related to A?
2. What is related to G?
3. What space is related to D?
4. What attaches within C?

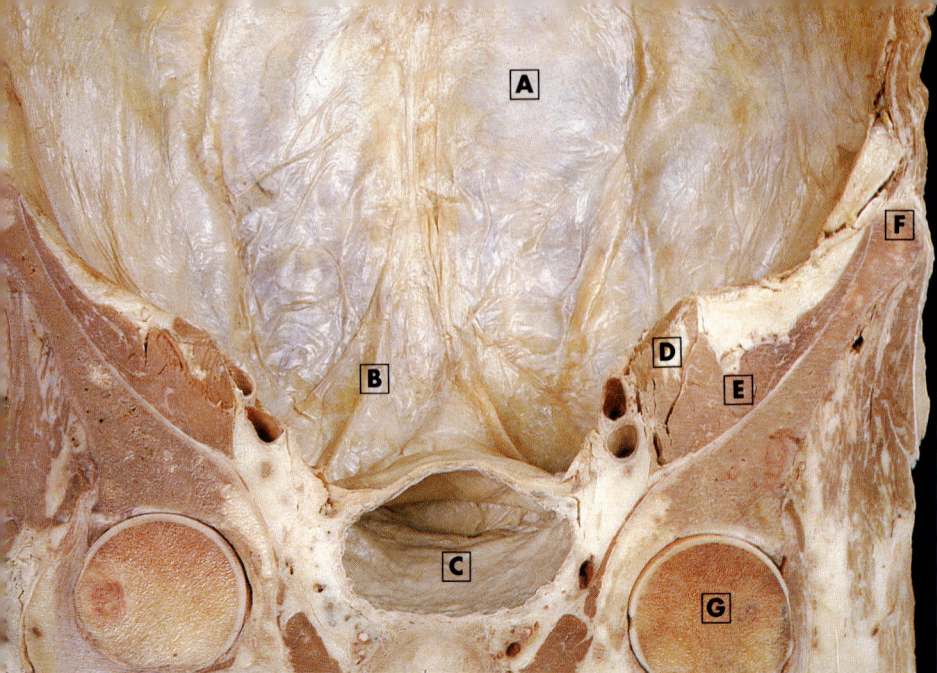

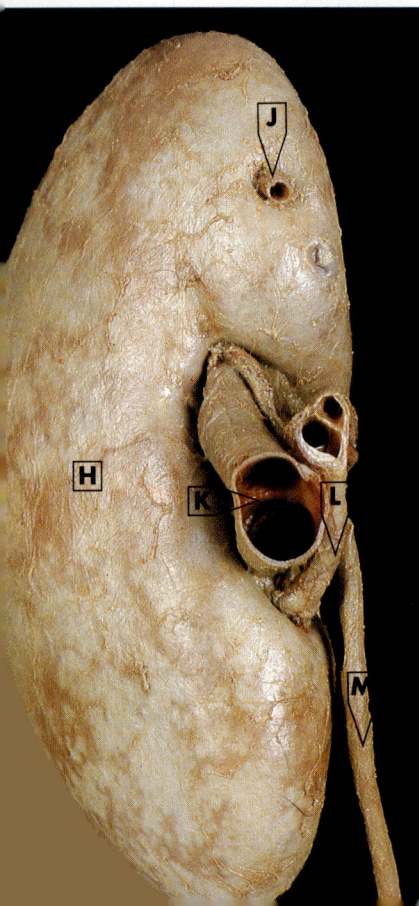

1. What is the nerve supply to A?
2. What is the embryological significance of B?
3. What is the nerve supply to D?
4. What is the action of E?
5. What parts of the intestine are related to H?

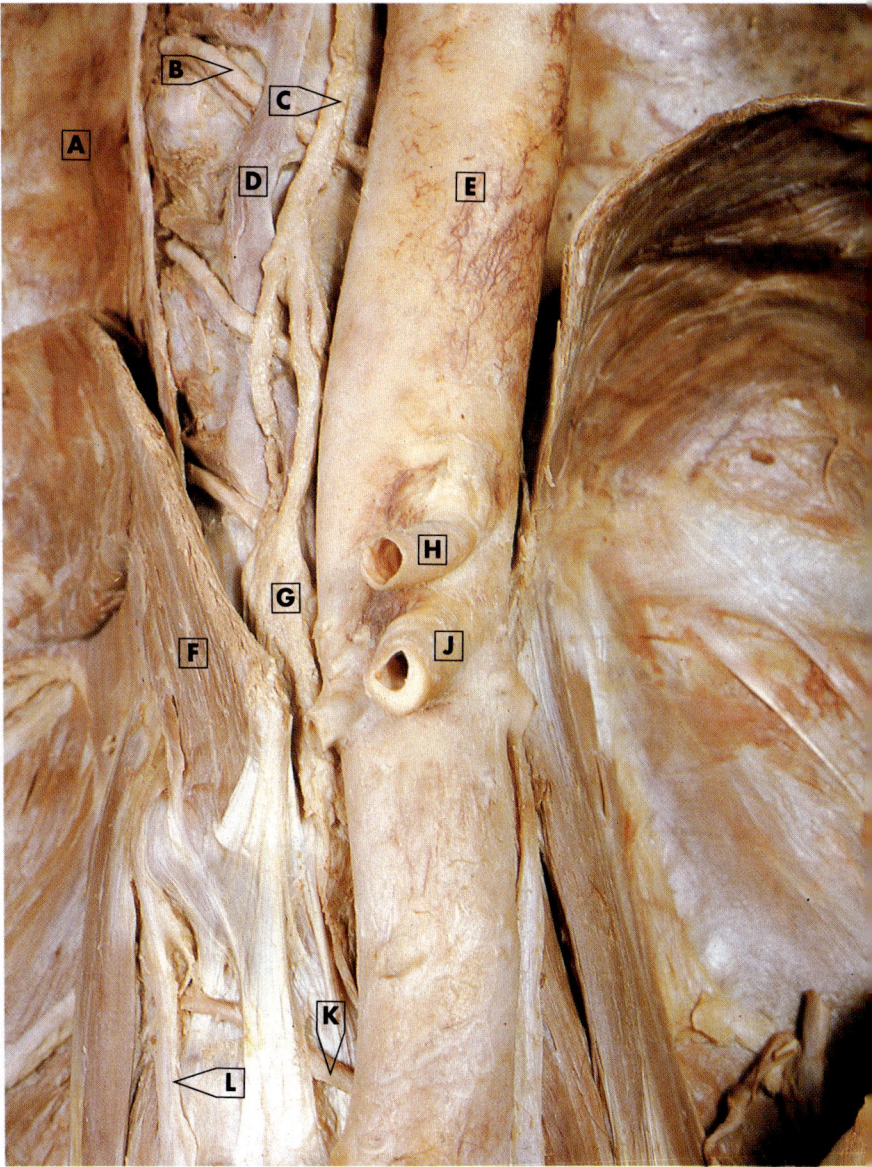

1. What is the termination of C?
2. What is the termination of D?
3. What are the branches of H?
4. What is the distribution of J?

61

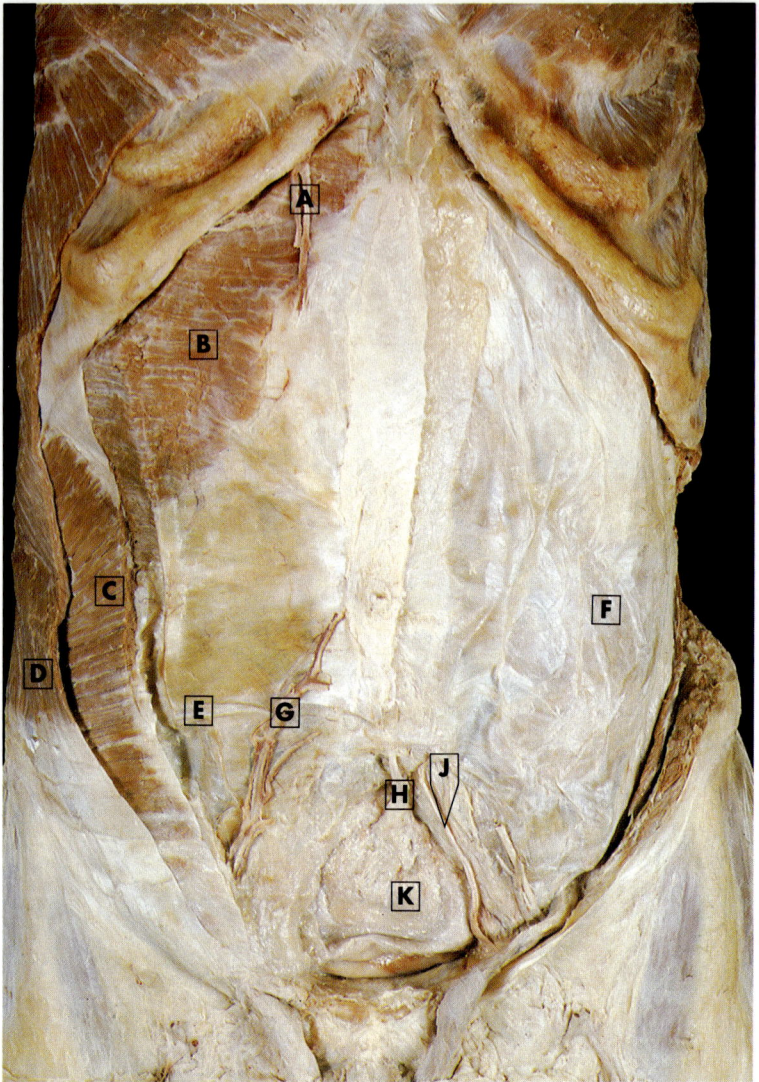

1. What lies anterior to A?
2. What is the origin of the artery at G?
3. What is the embryological significance of H?

4. Which of the muscles B, C and D rotate the lumbar spine?

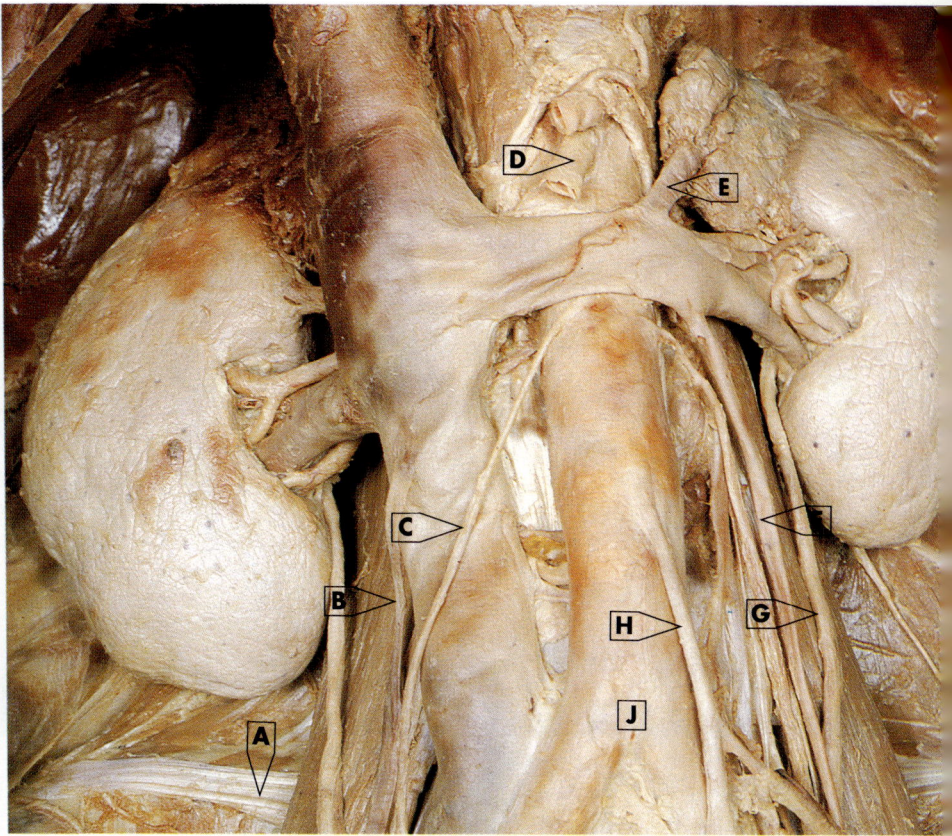

1. Are C and H visceral or parietal branches of the aorta?
2. What name is given to the distal part of H?
3. At what vertebral level is J?

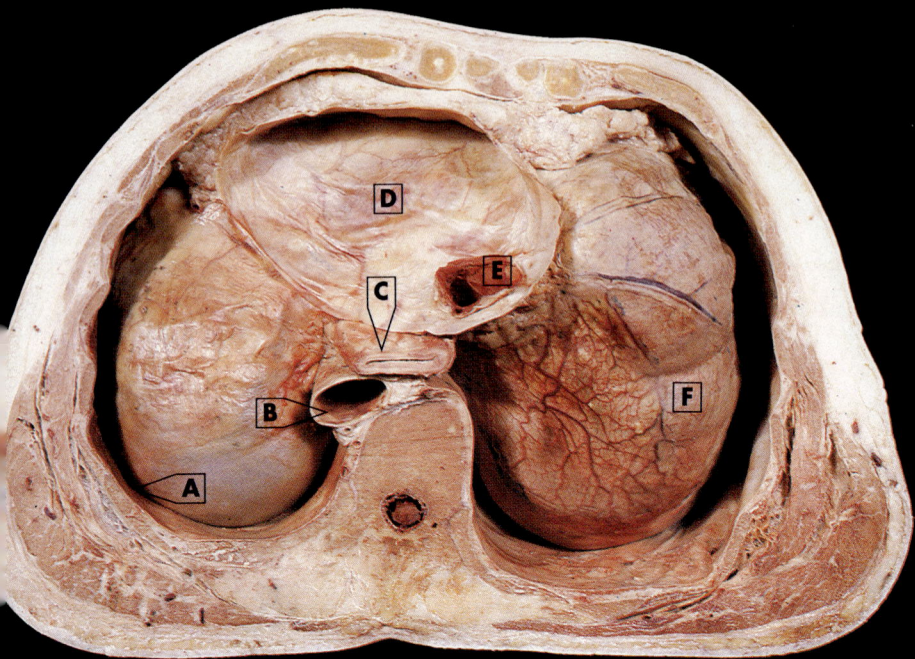

1. What is the nerve supply to C?

2. Which chamber of the heart lies anterior to C?

3. What lies inferior to F?

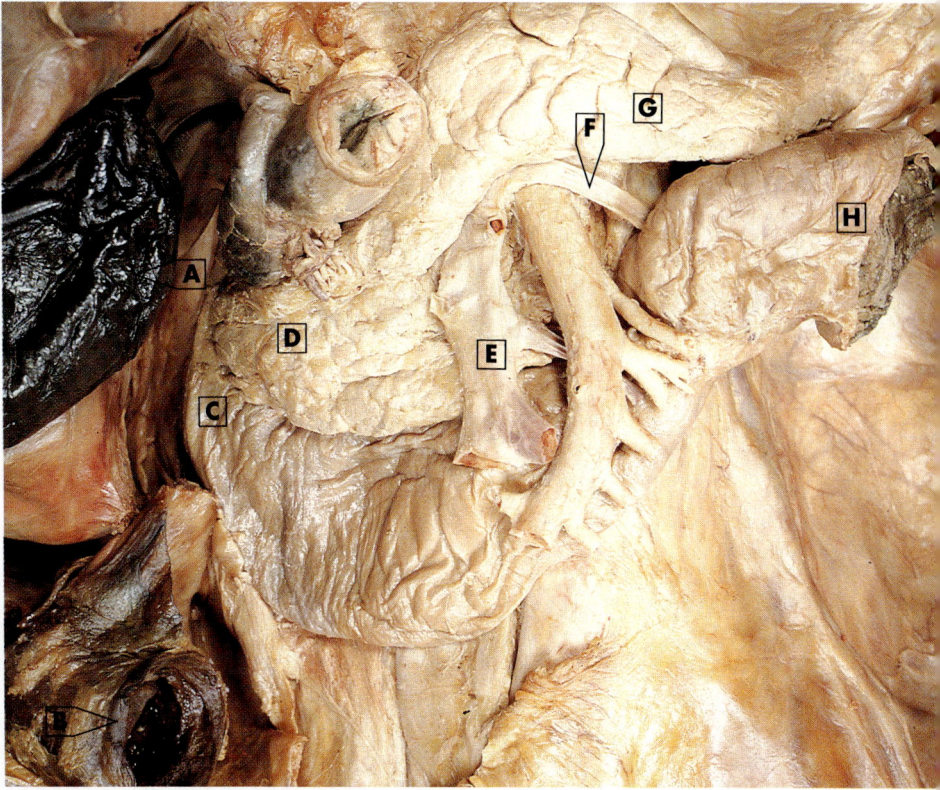

1. Which artery supplies A?
2. What lies posterior to E?
3. Which organs does F drain?

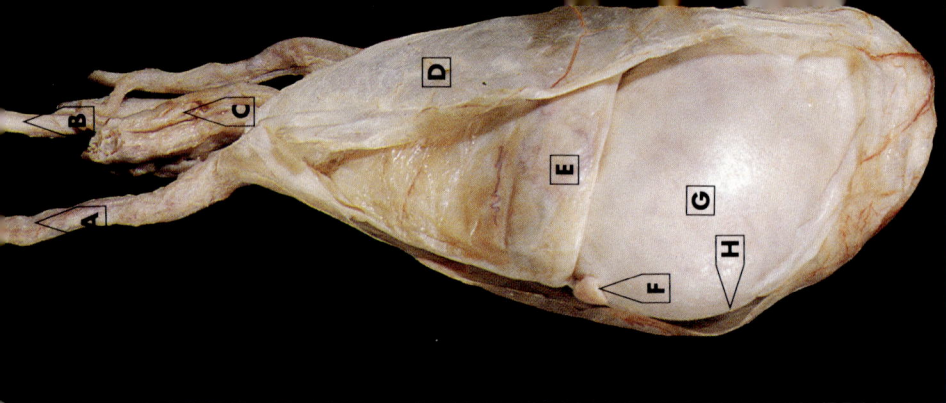

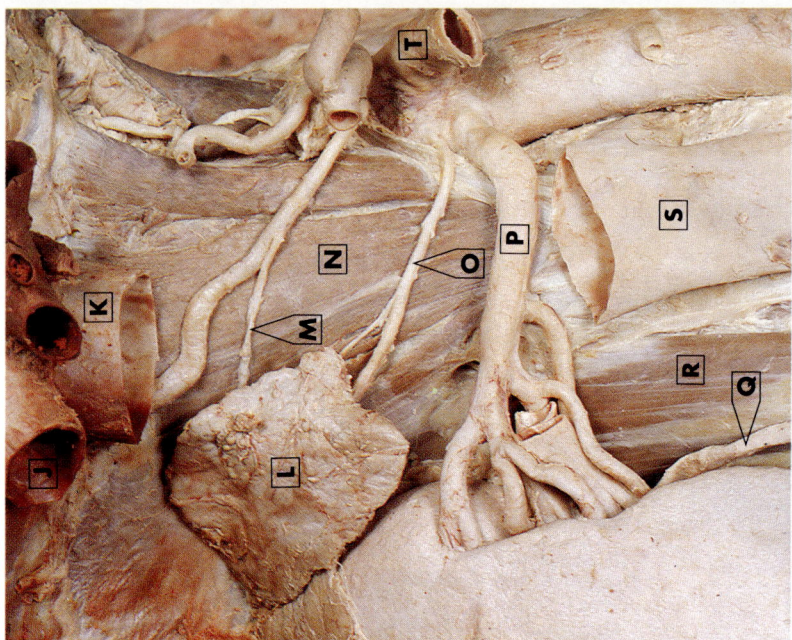

1. What is the origin of A?

2. What is the termination of B?

3. What is the clinical term for an accumulation of serous fluid within H?

4. What lies anterior to L?

5. What lies anterior to P?

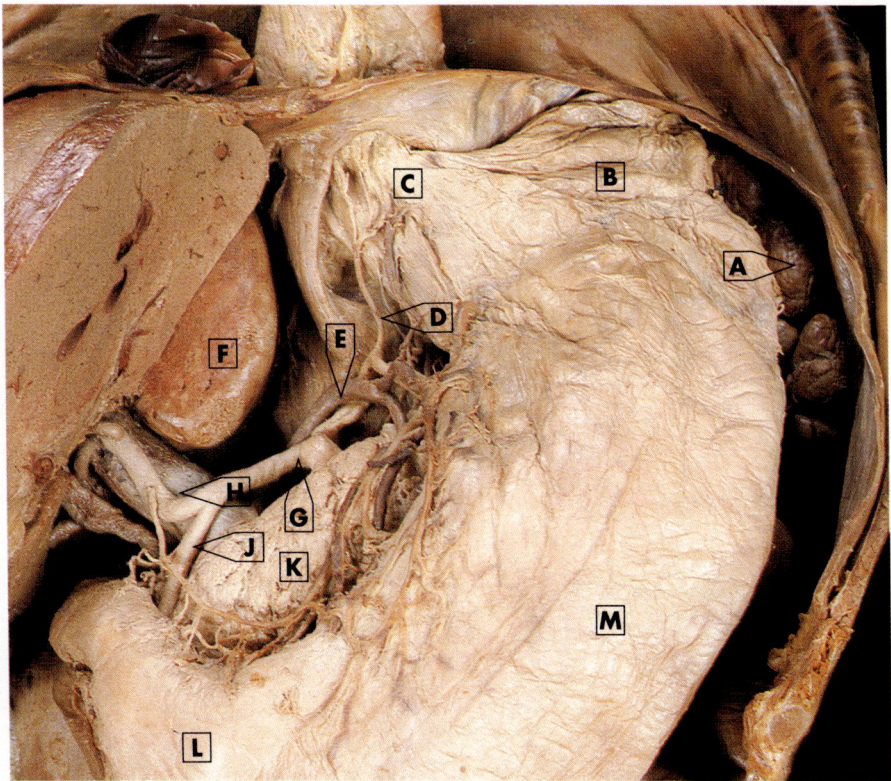

1. What is the termination of E?
2. What are the branches of J?
3. What lies anterior to K?
4. What lies anterior to M?
5. What lies posterior to H?

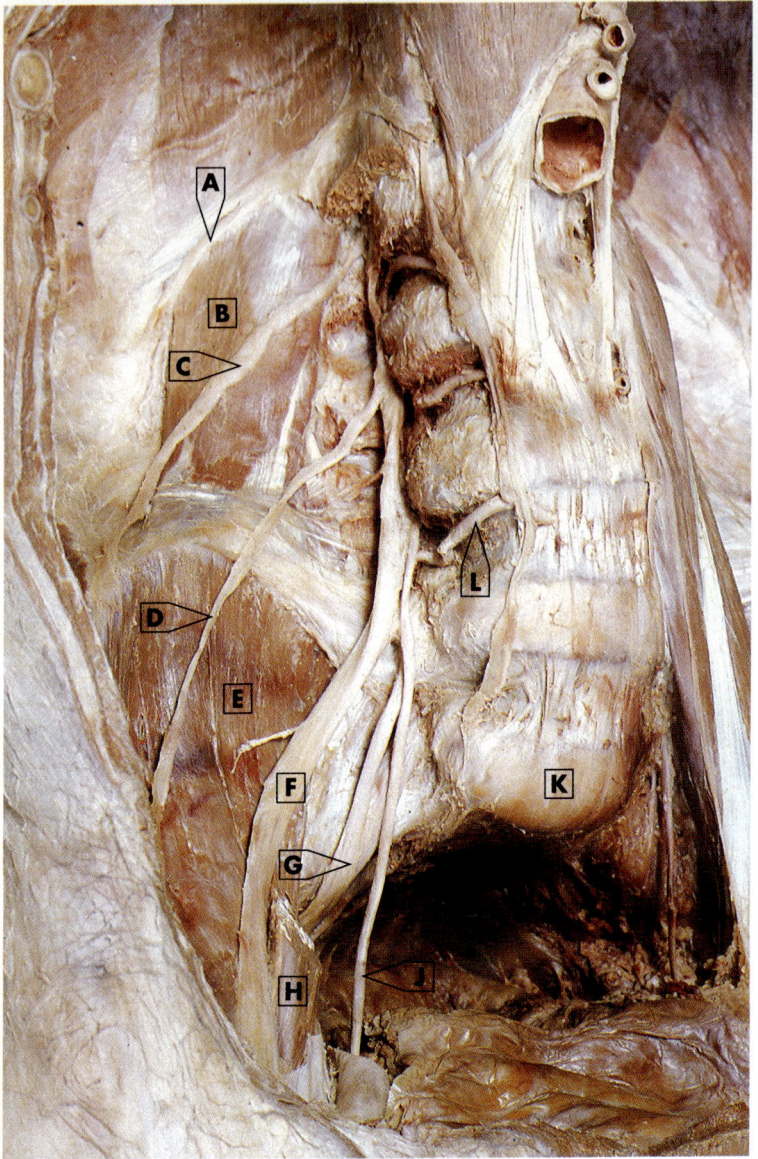

1. What is the action of B?
2. What is the segmental origin of G?
3. What are the actions of H?

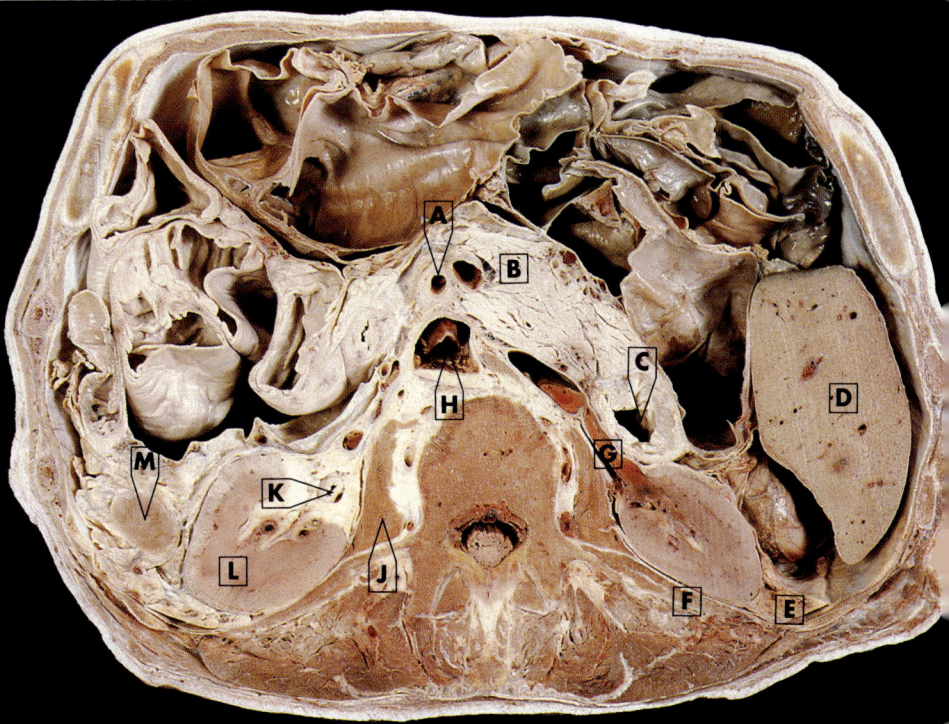

1. What is the origin of A?
2. Name the tissue that surrounds E.
3. What is the termination of K?
4. Is M palpable in a healthy living subject?

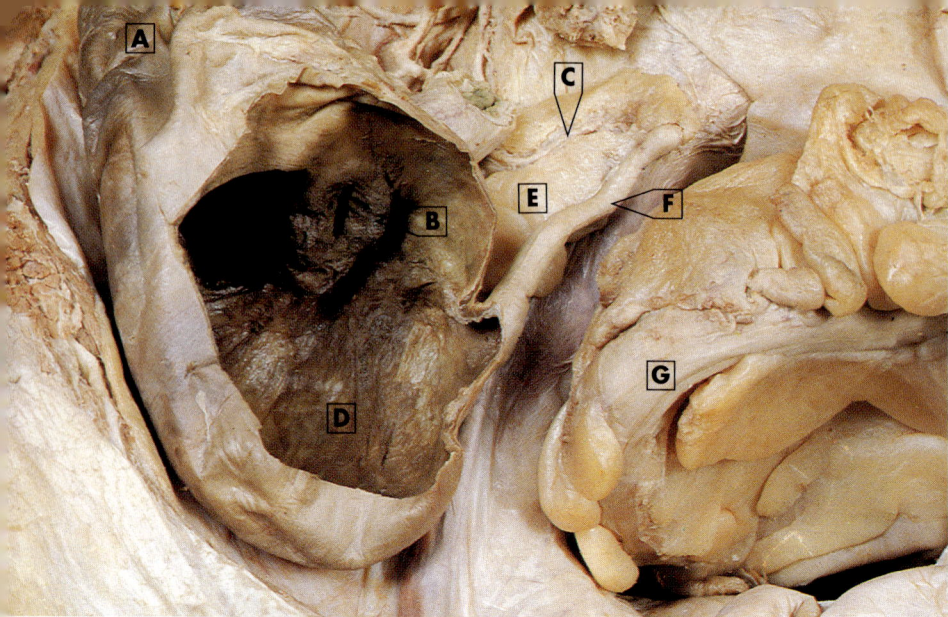

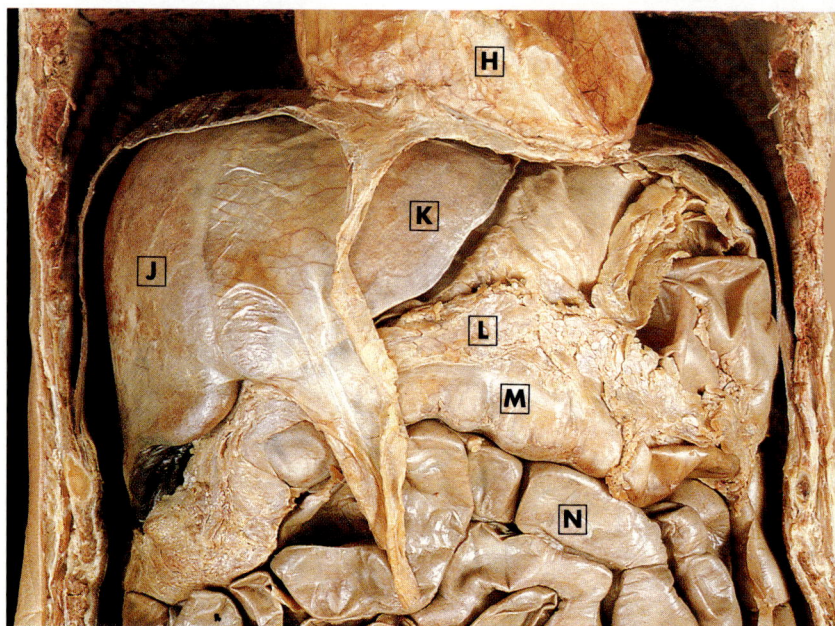

1. What is the origin of C?
2. What is the termination of G?
3. What lies deep to H?
4. In what way is L in this specimen unusual?

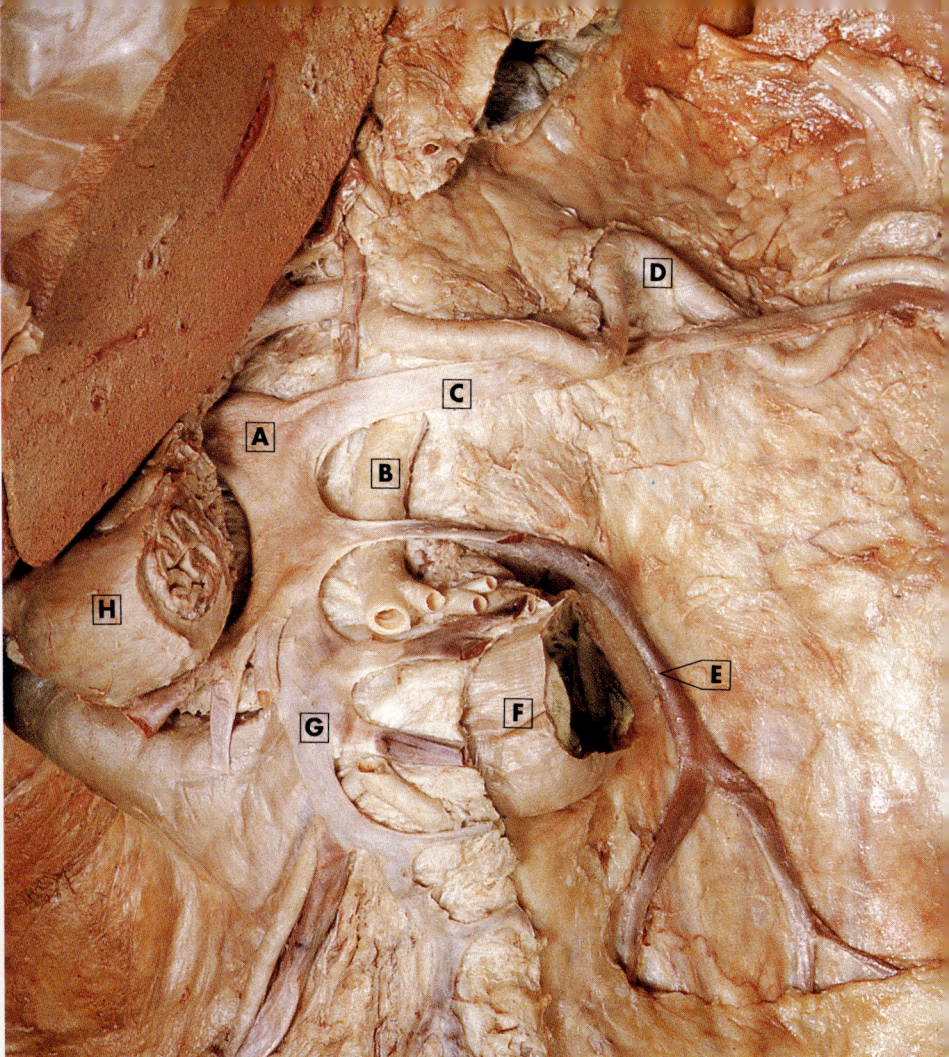

1. What is the termination of A?
2. What is the distribution of D?
3. Comment on the termination of E.
4. What lies posterior to F?

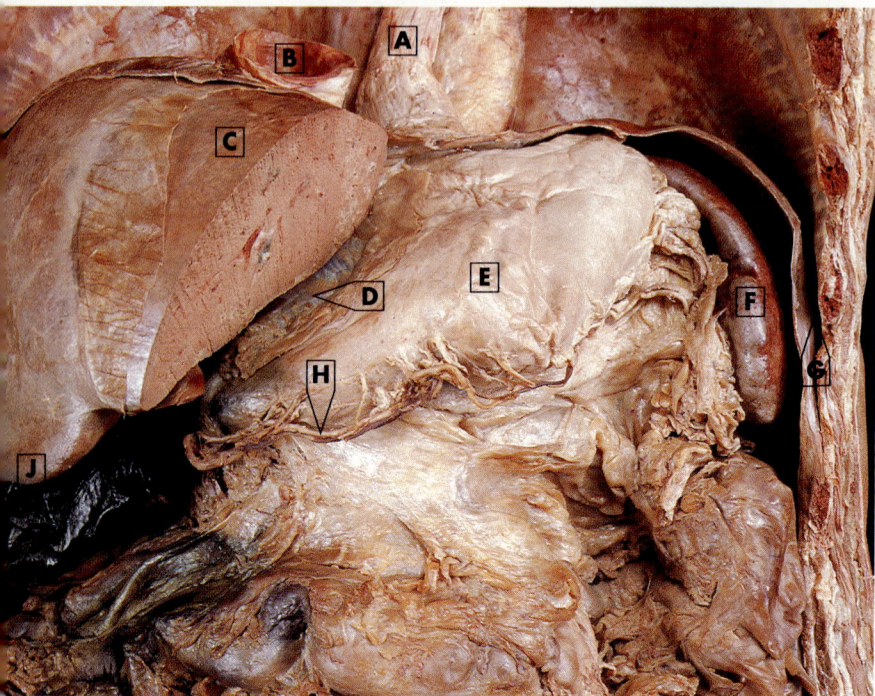

1. What is the nerve supply to A and to E?
2. What vessels lie within D?
3. What is the distribution of L?
4. What is the origin of P?

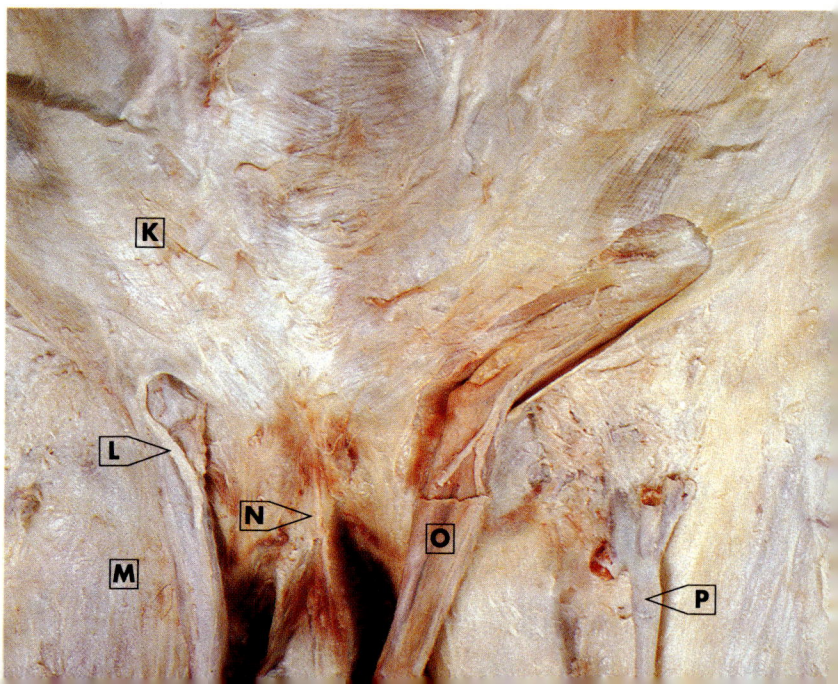

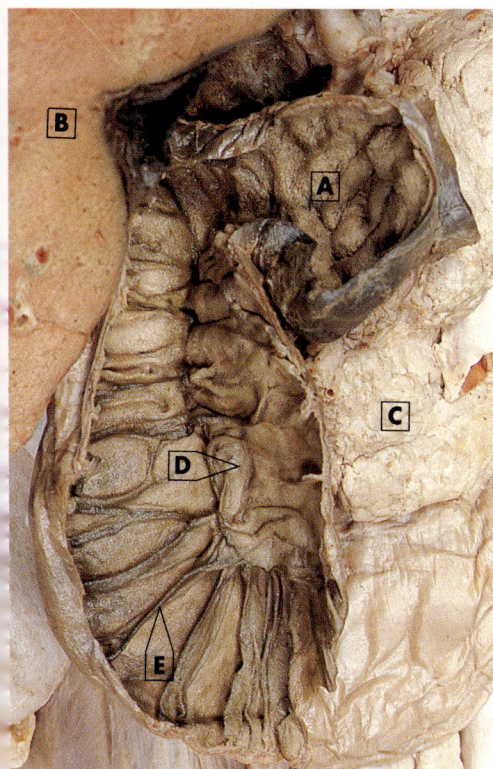

1. In clinical practice, what alternative names may be given to A?
2. Name the ducts that terminate at D.
3. What is the segmental origin of K?
4. Name the aperture between H and F.

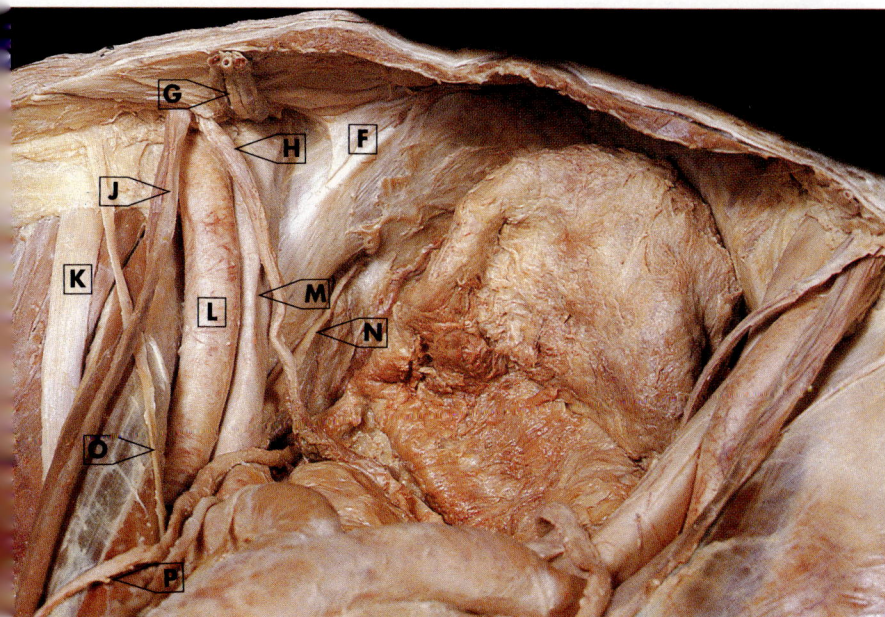

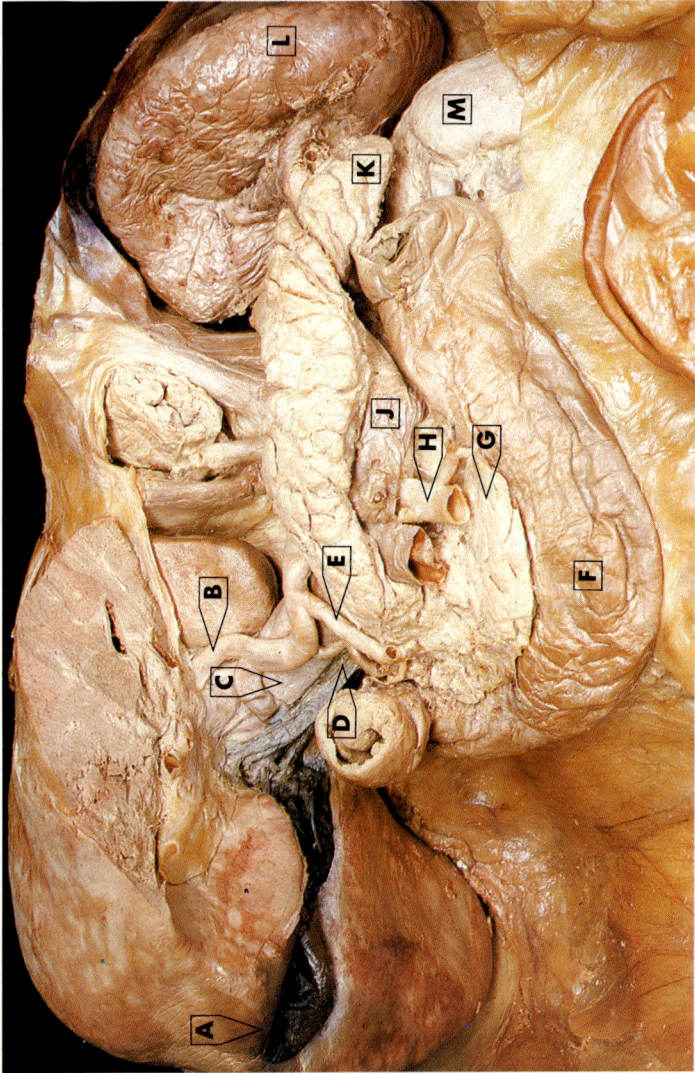

1. What is the surface marking of A?

2. What is the distribution of B?

3. What lies anterior to E?

4. What is the relationship of K to the peritoneum?

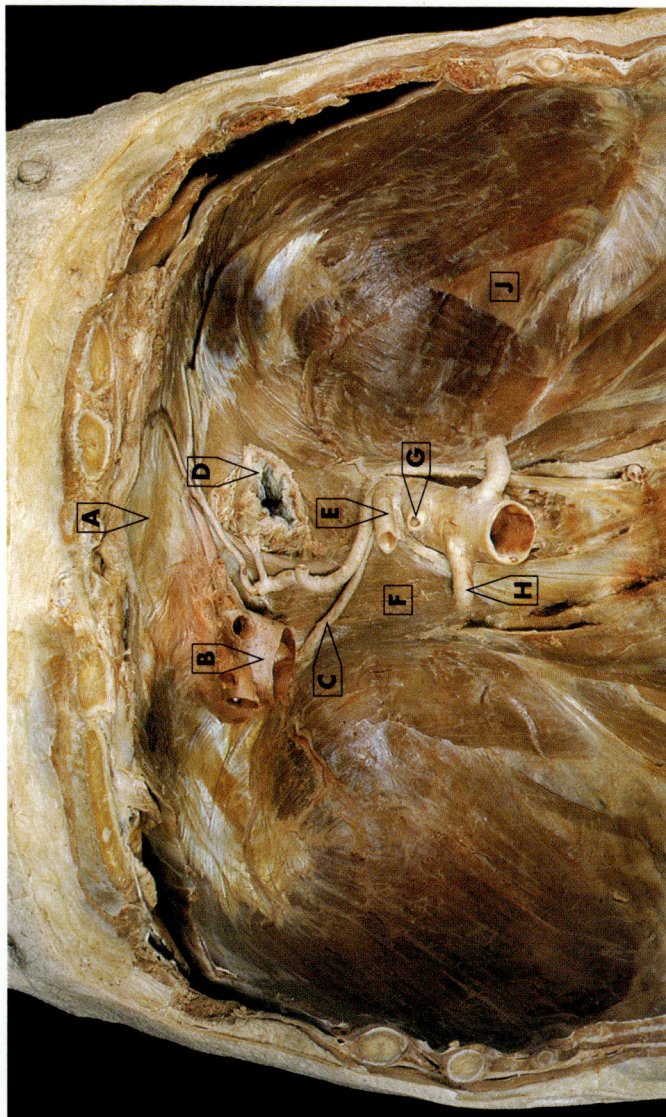

1. What lies superior to A?

2. What is the origin of B?

3. What structures accompany D from thorax to abdomen?

4. What is the motor innervation to F?

75

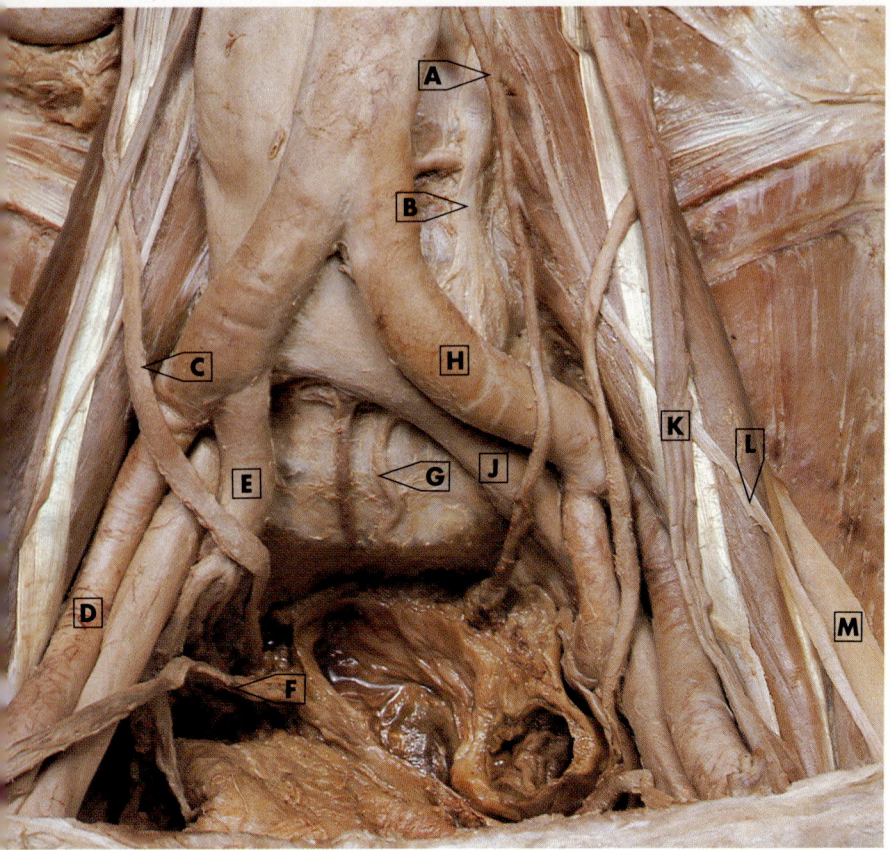

1. What is the embryological significance of A?
2. What is the termination of D?
3. Name a branch of D.
4. What is the distribution of E?

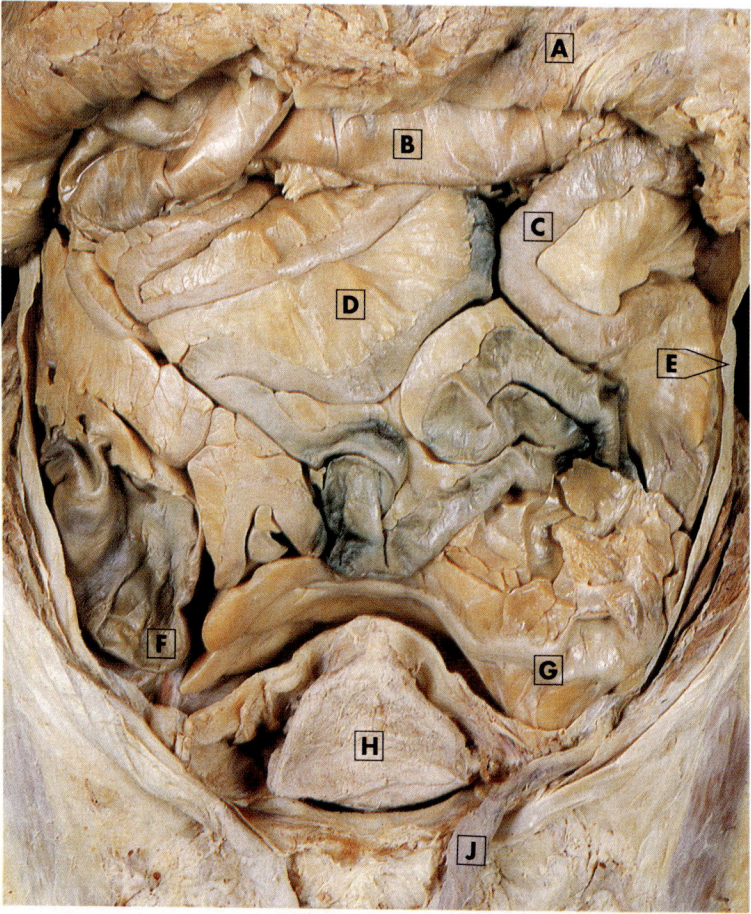

1. What is A attached to?
2. What vessels lie within D?
3. What is the nerve supply to E?
4. What distinguishes G from C?

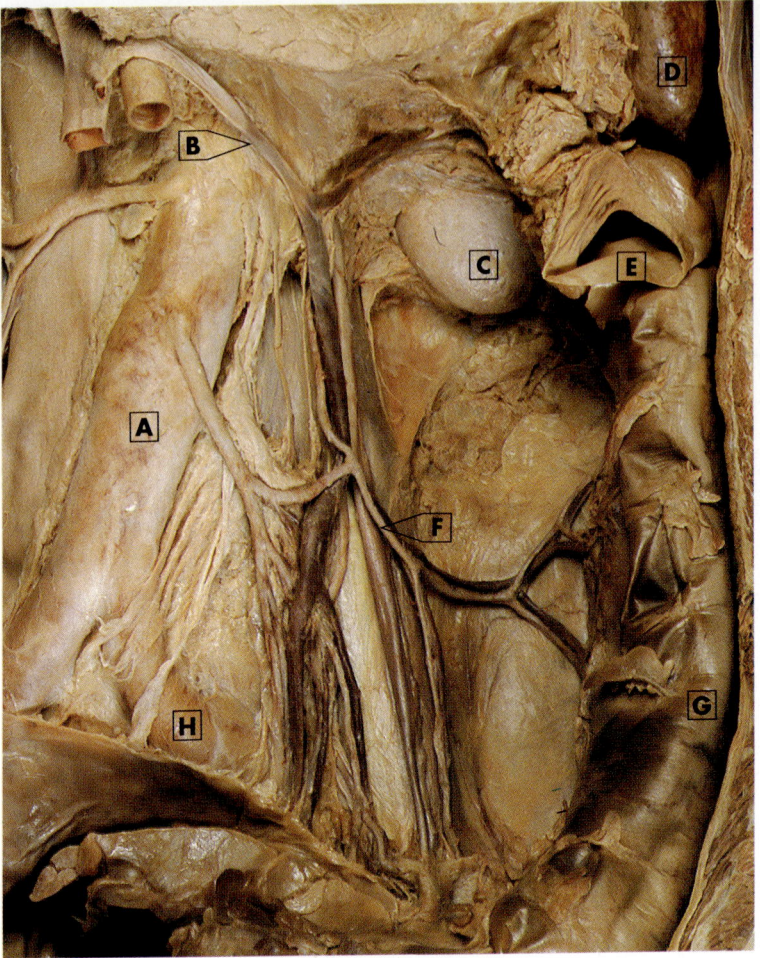

1. Name the parietal branches of A.
2. What are the anterior relations of C?
3. Is G palpable in the living subject?
4. What is the termination of H?

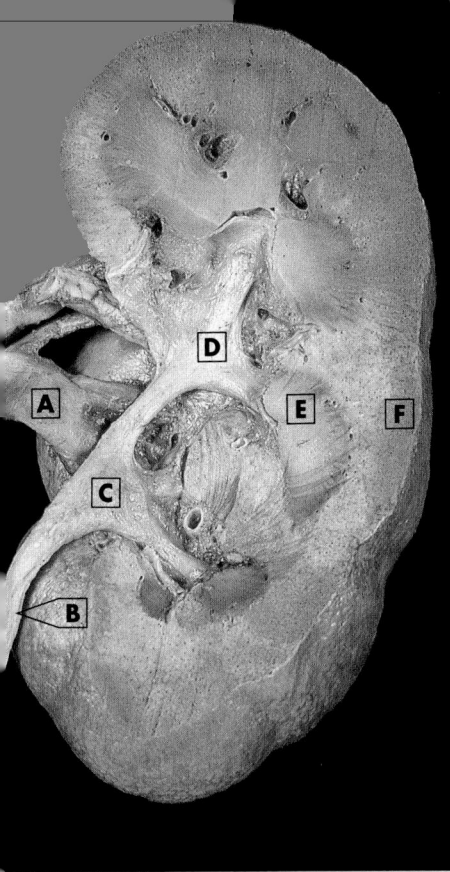

1. To which organs is the abdominal part of B related?
2. What are the relationships of M to the duodenum?
3. What are the relationships of N to the peritoneum?
4. What is the embryological significance of O?

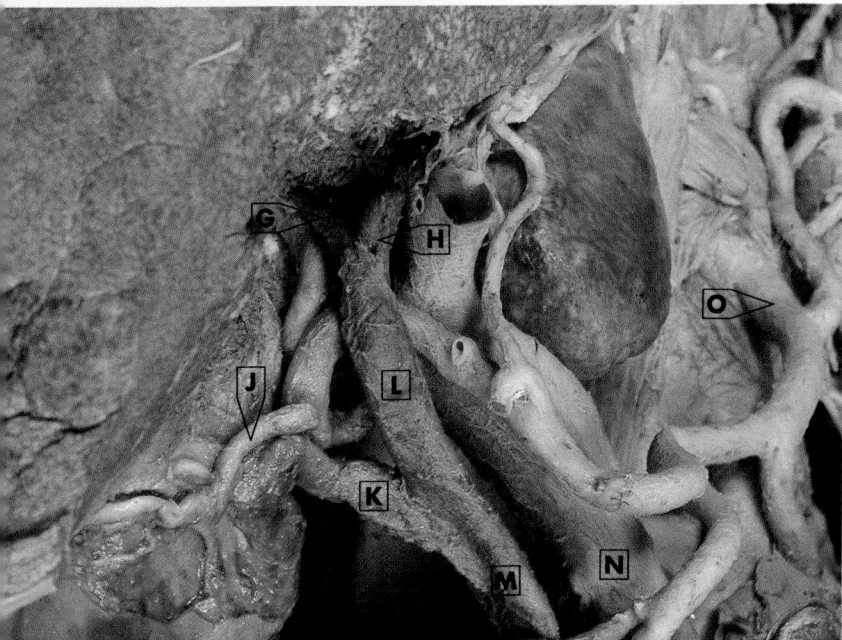

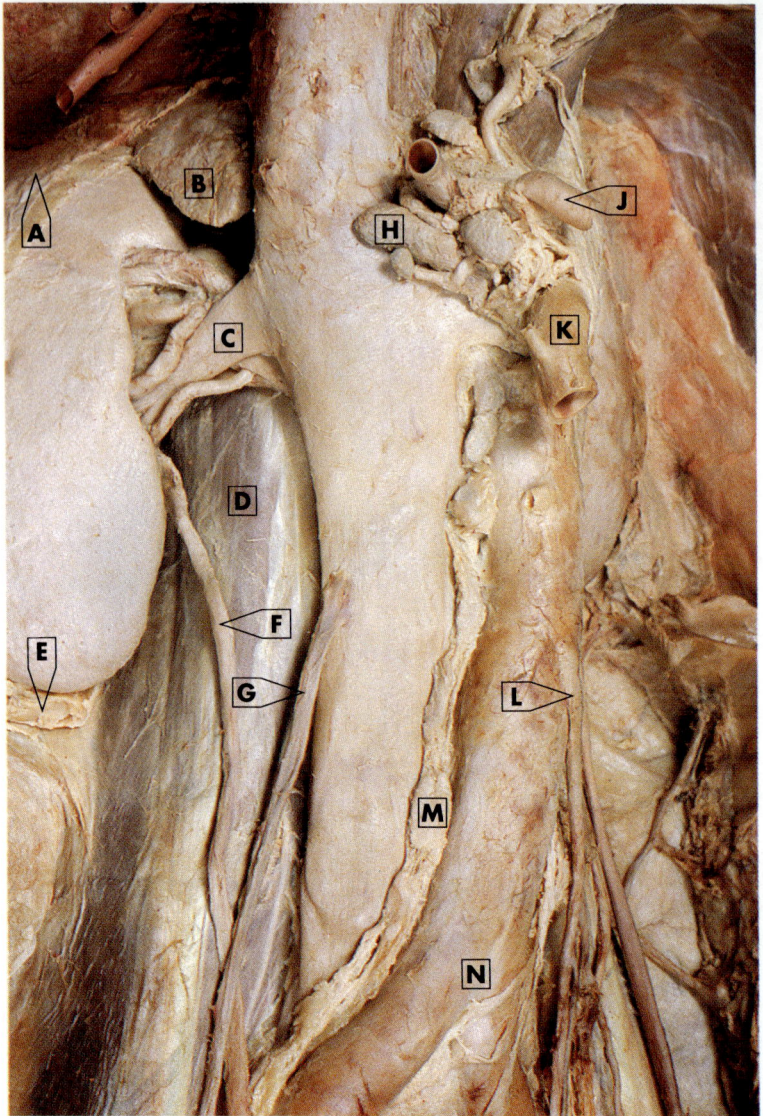

1. What lies posterior to the proximal portion of K?
2. Which part of the gut commonly overlies the origin of L?
3. Lymph from H drains to where?

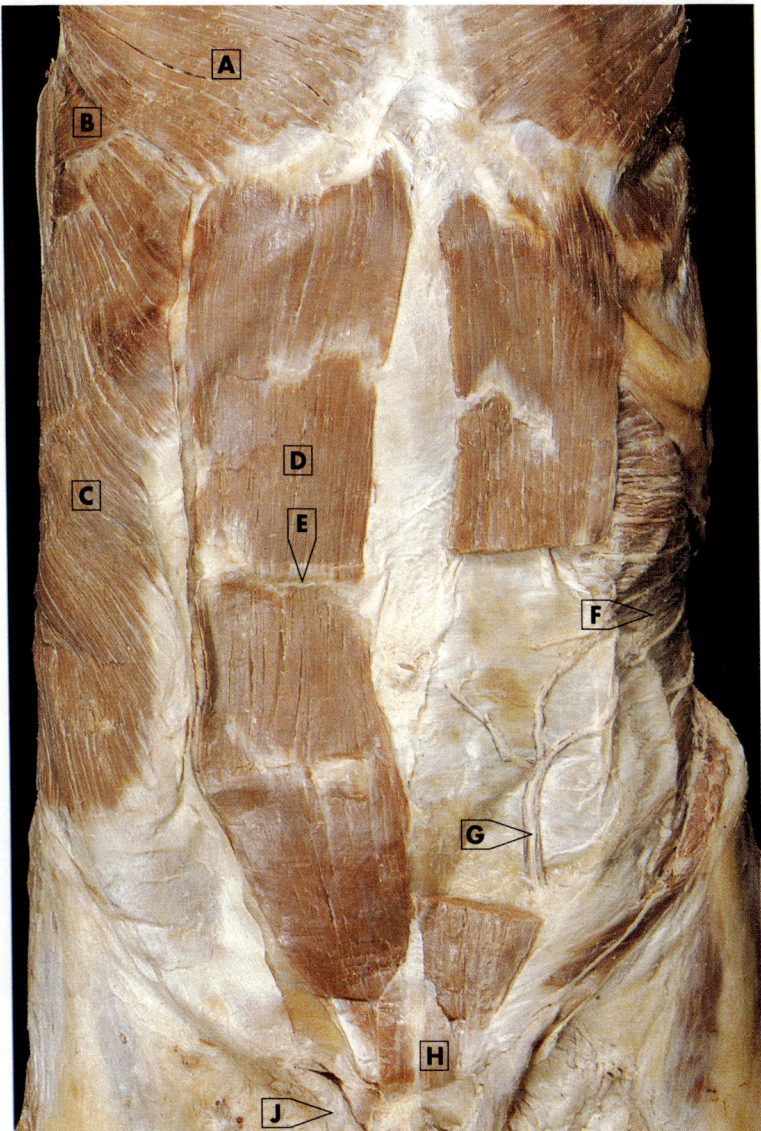

1. What are the actions of D?
2. Is the anterior layer of the rectus sheath attached to E?
3. What is the origin of G?
4. What structures lie within J?

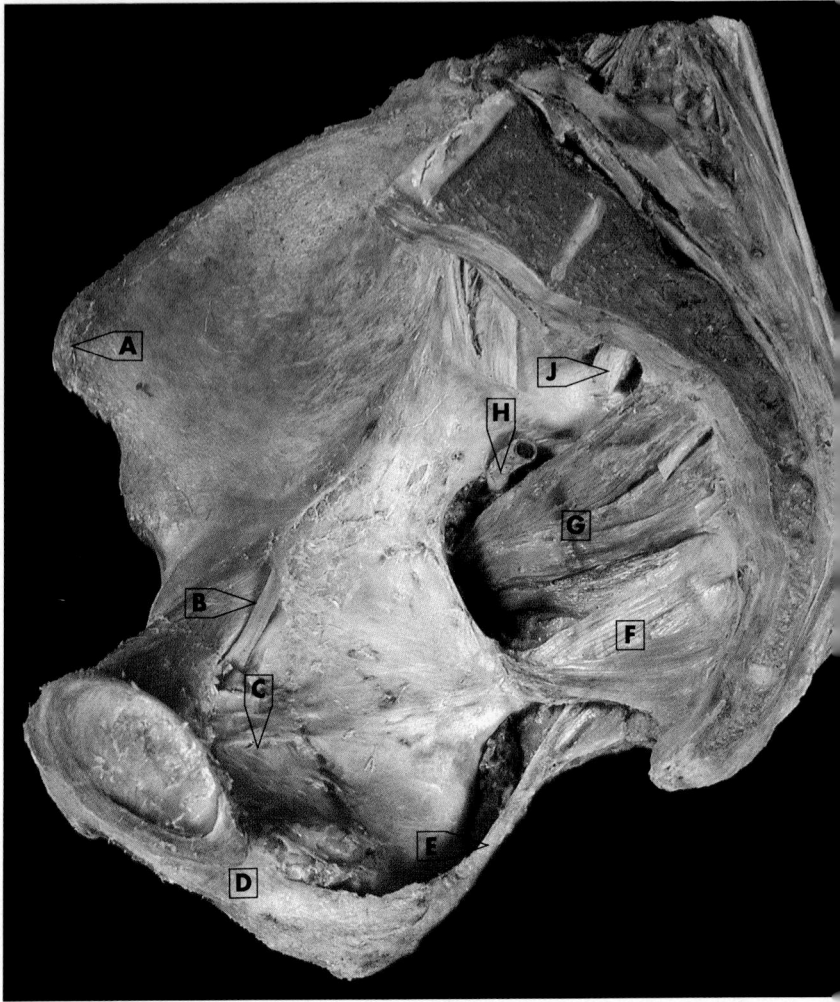

1. What aperture does G traverse?
2. What is the distribution of B?
3. What is related to C?

4. What is the origin of H?

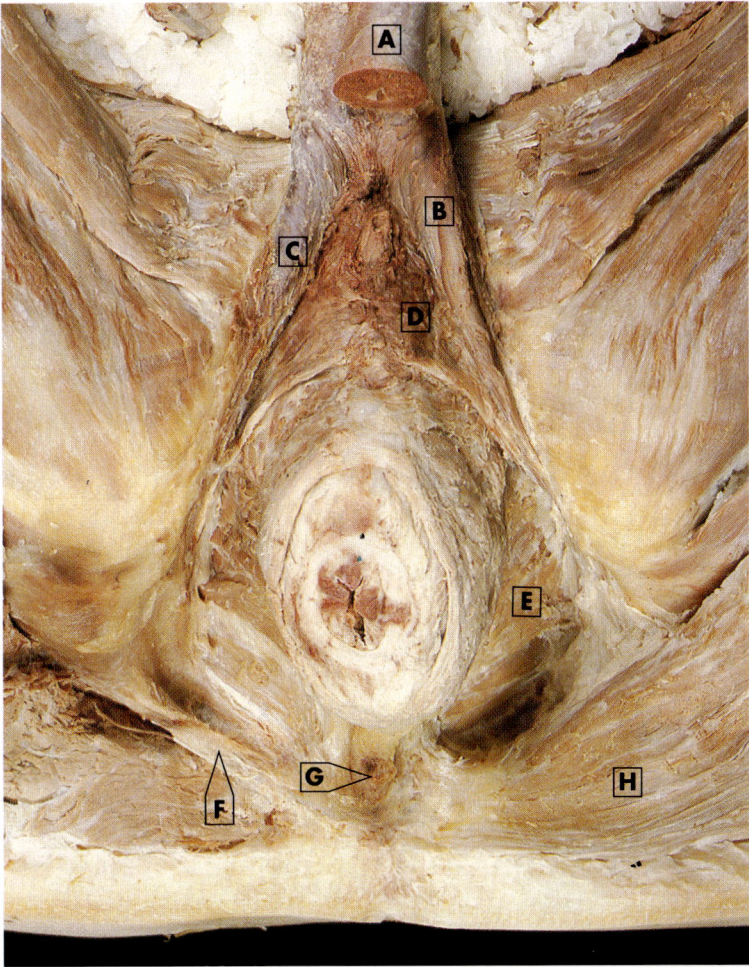

1. How does A terminate distally?
2. What is the nerve supply to B?
3. What lies superior to D?
4. What lies inferior to E?

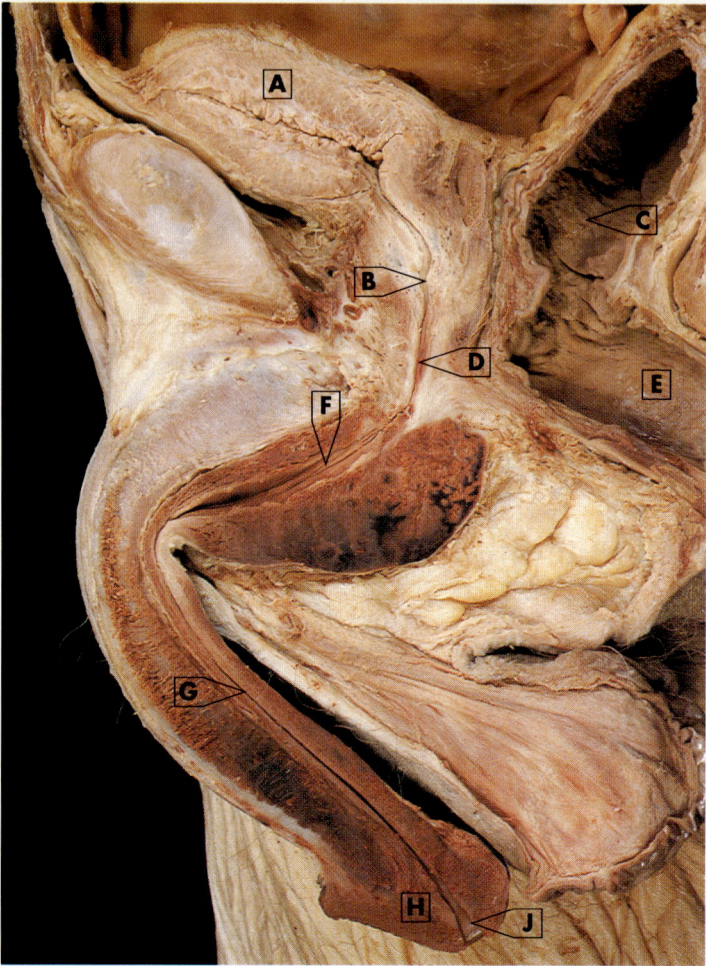

1. What is the motor nerve supply to A?

2. What muscles are related to D?

3. What muscles are related to E?

4. What is the sensory nerve supply to G?

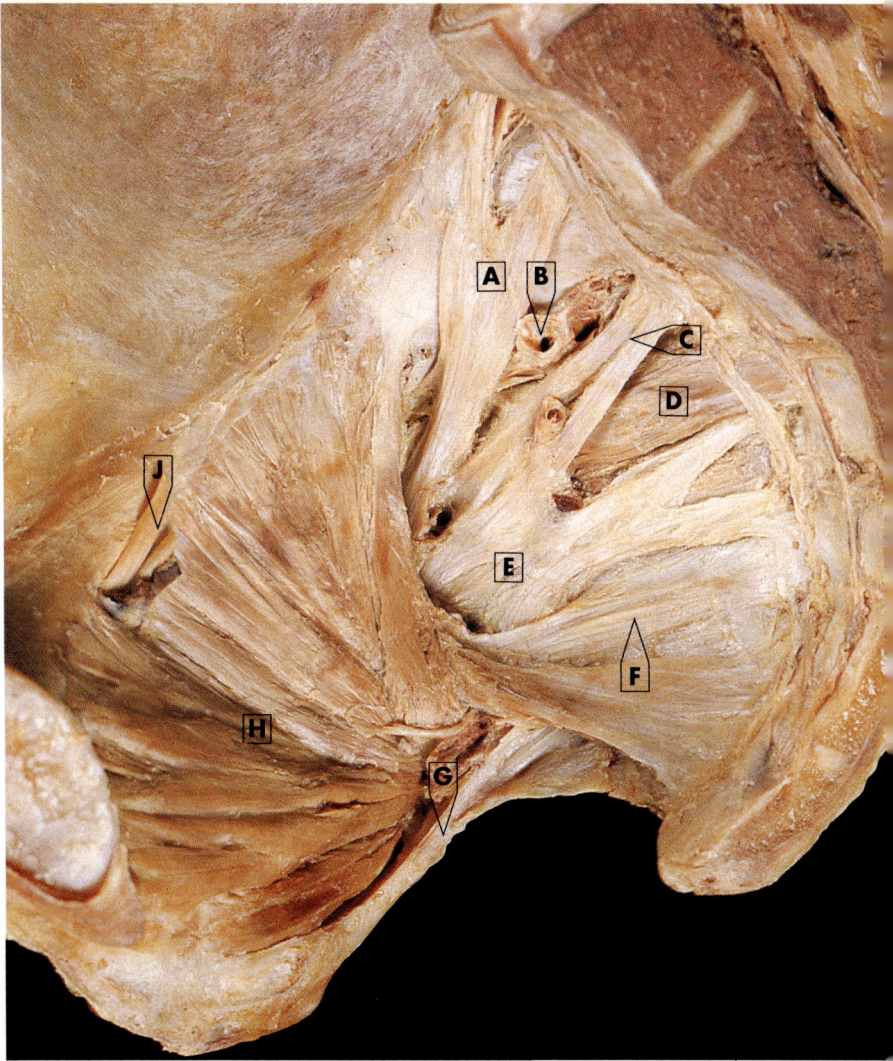

1. What is the segmental origin of A?
2. Which joint is stabilized by F and G?
3. What is the action of H?
4. What is the origin of the artery at J?

85

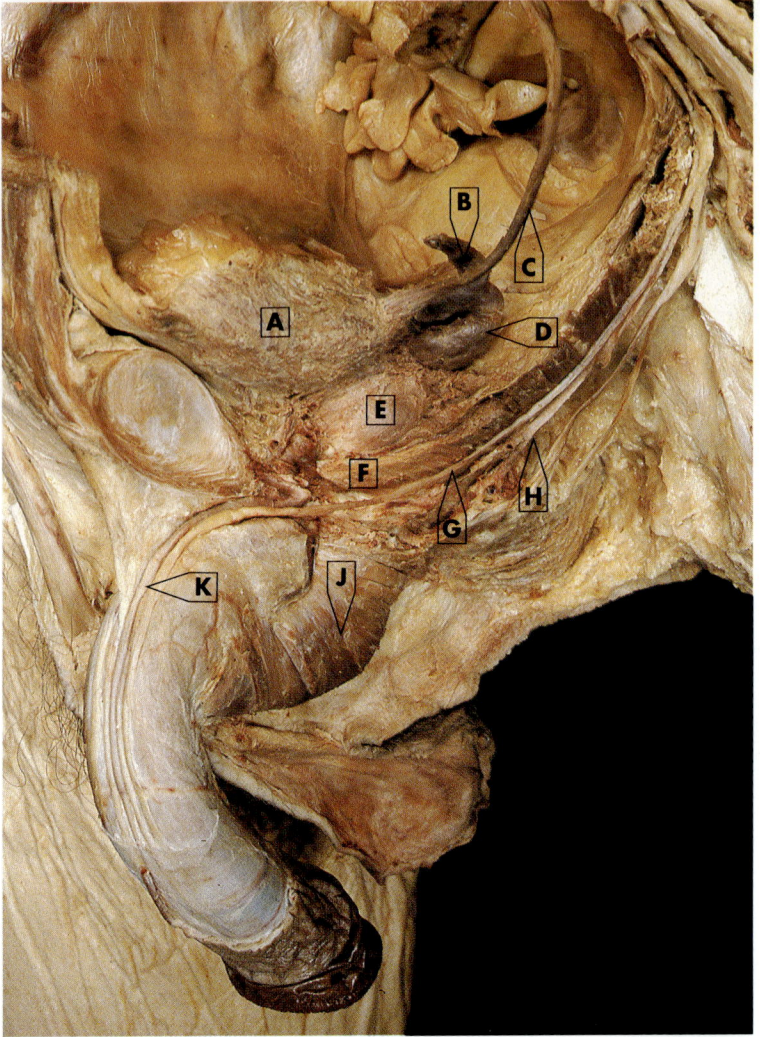

1. What is the distribution of G?
2. What is the origin of H?
3. What is the nerve supply to J?

4. What is the termination of K?

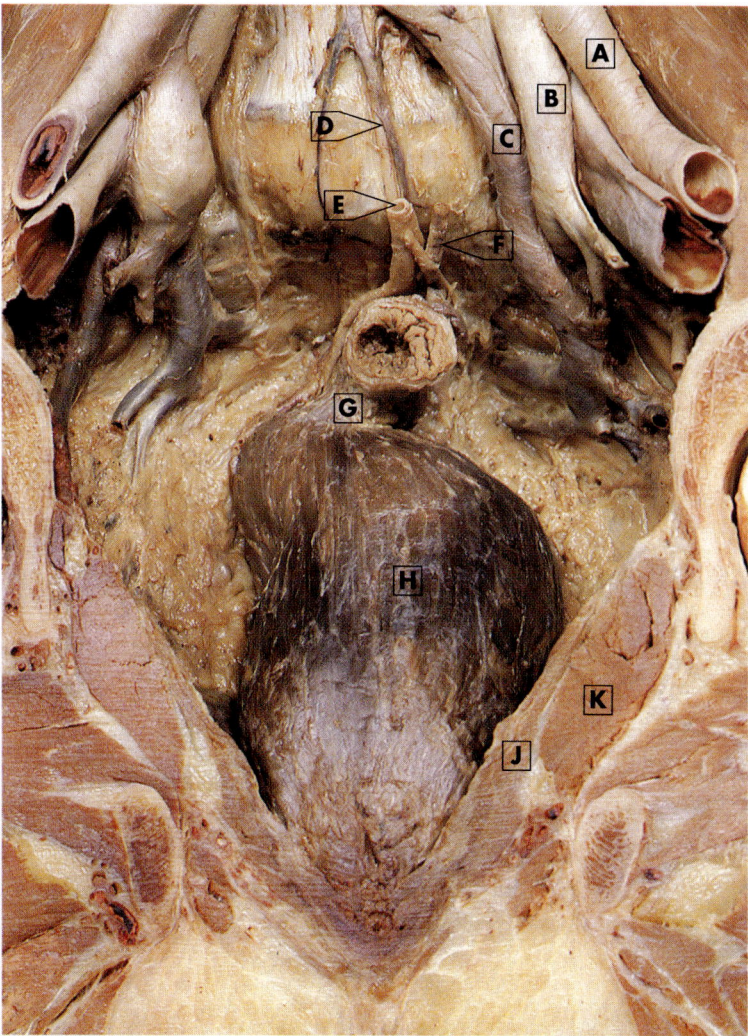

1. What is the origin of the artery at D?
2. Which of the veins visible in the dissection are portal and which are systemic?
3. What are the anterior relations of H in the male?
4. What is the nerve supply to J?

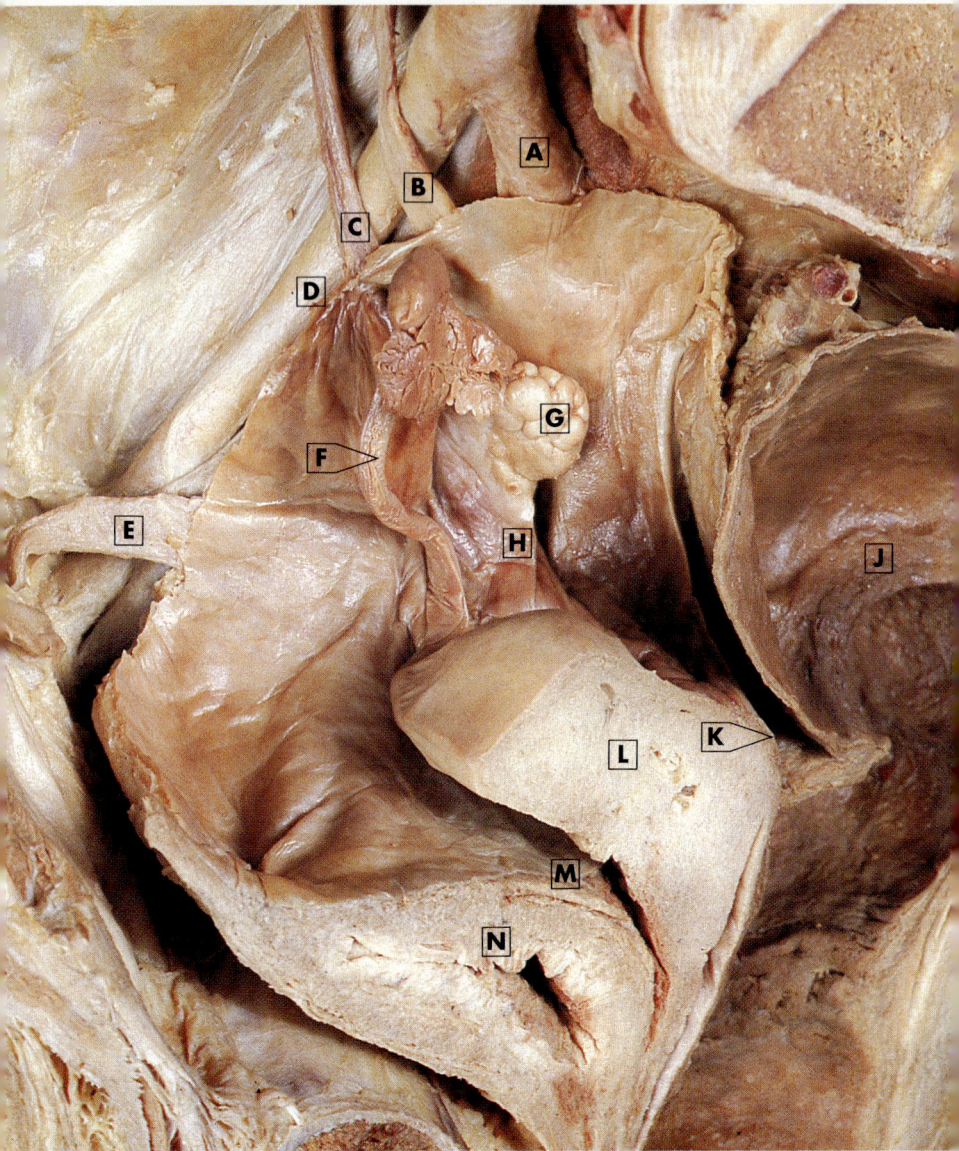

1. What is the termination of E?
2. What arteries accompany F?
3. Name the perioneal fold that encloses E, F and H.

4. In clinical practice what eponymous term is often applied to K?

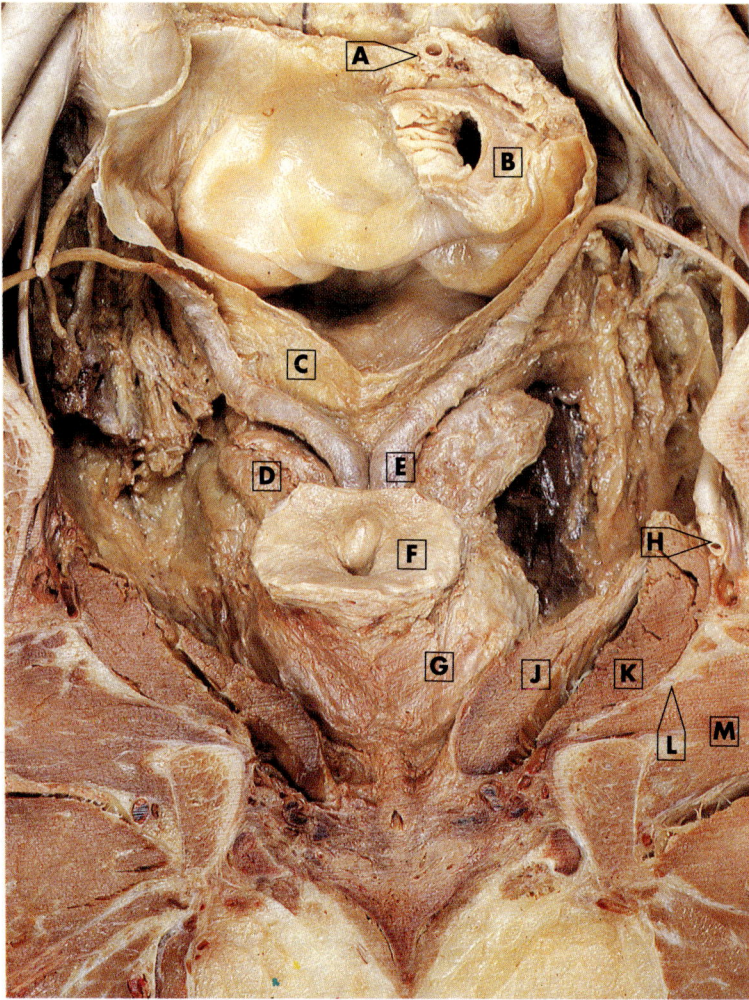

1. What is the termination of E?
2. Contrast the sphincters lying superior and inferior to G.
3. What is the distribution of H?

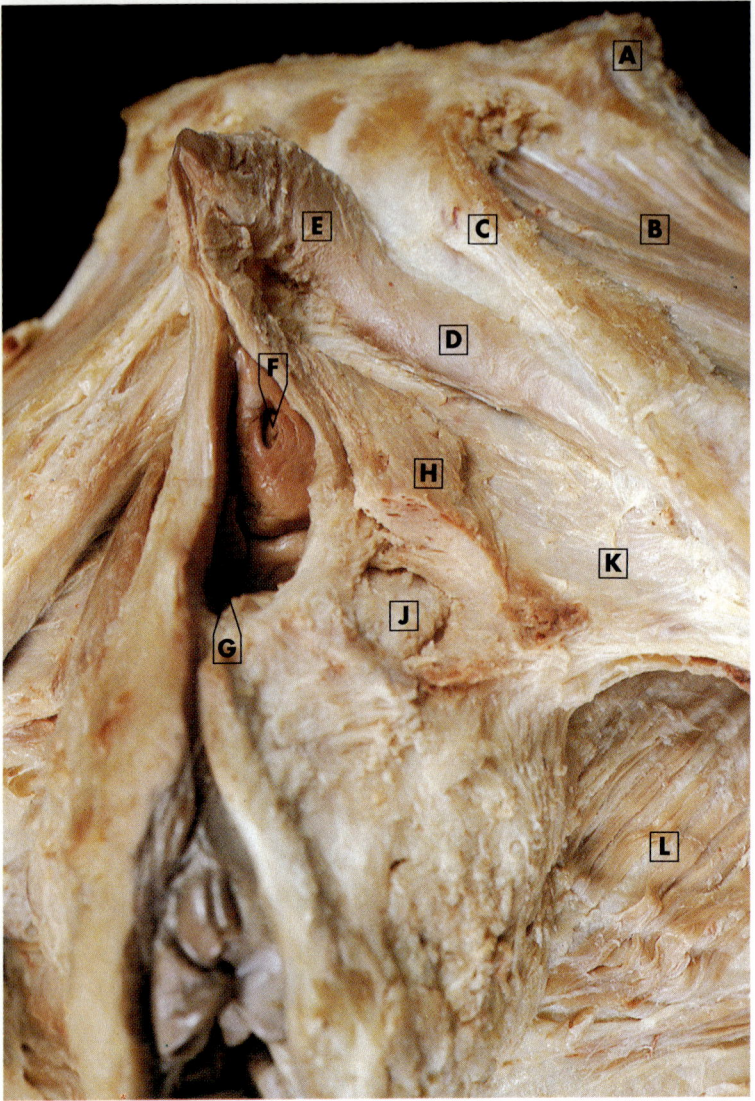

1. What lies superficial to D?
2. What lies superficial to H?
3. What is the sensory nerve supply to F?
4. What is the eponymous name for J?

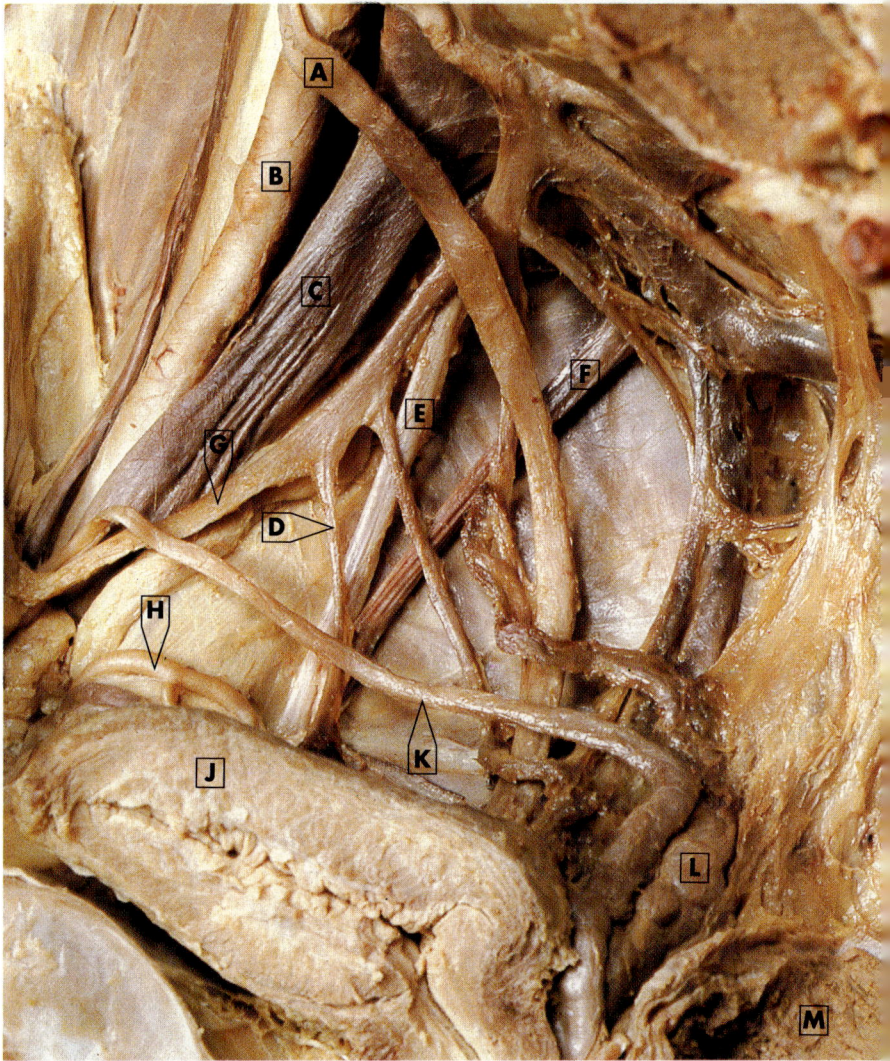

1. What is the segmental origin of E?
2. What is the termination of G?
3. What is the origin of H?
4. What is the origin of K?

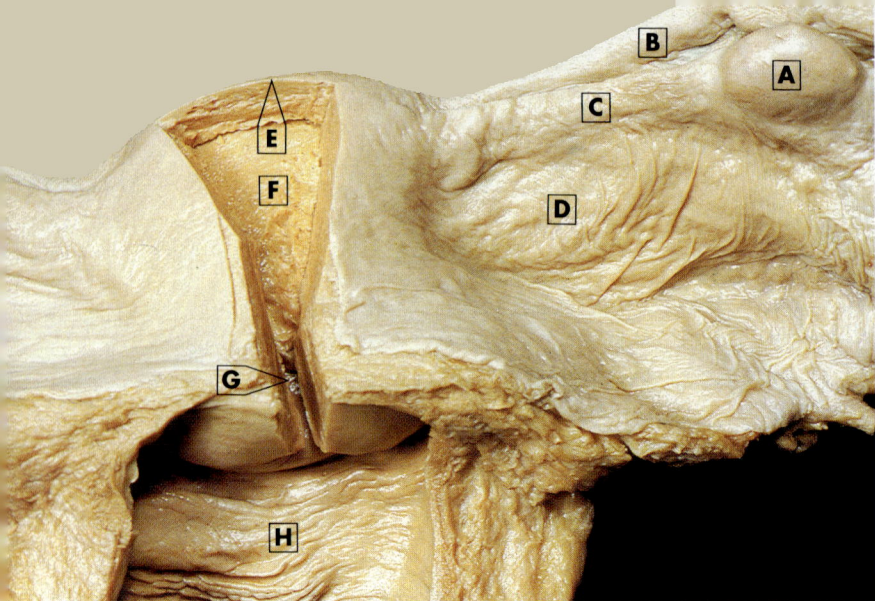

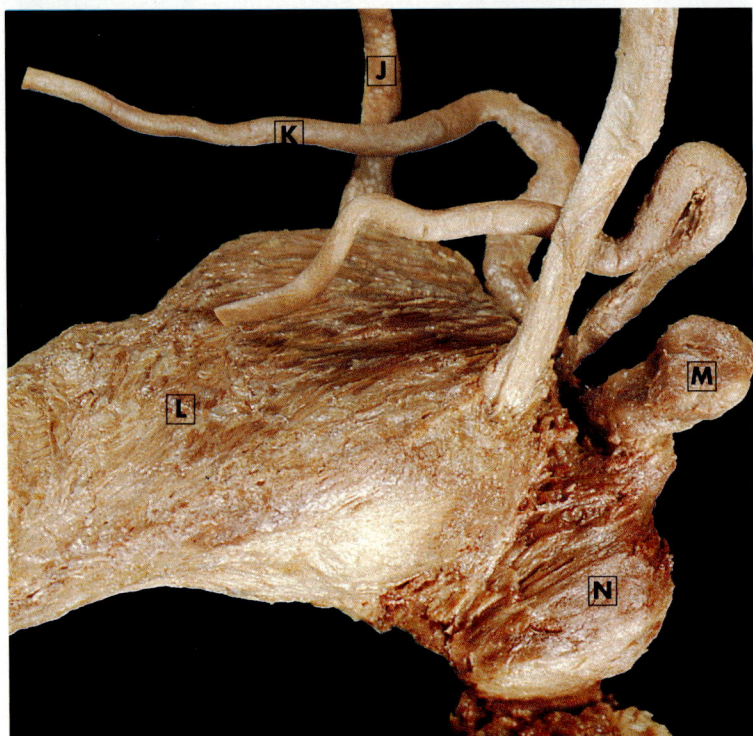

1. Name different parts of B.
2. What lies anterior to H?
3. What lies posterior to M?
4. What lies anterior to N?

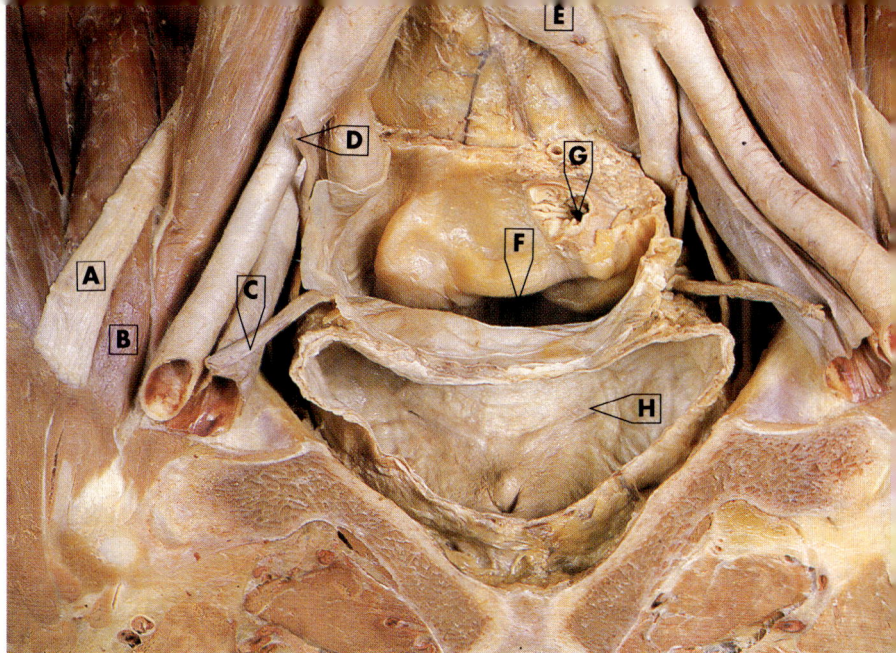

1. What is the segmental origin of A?
2. What is the termination of G?
3. What is the relationship of K to the uterus?
4. What is the embryological significance of Q?

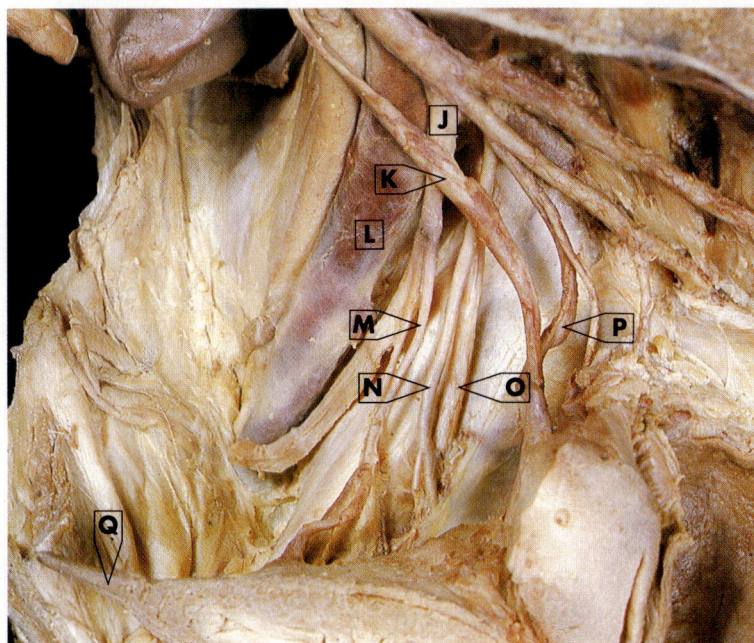

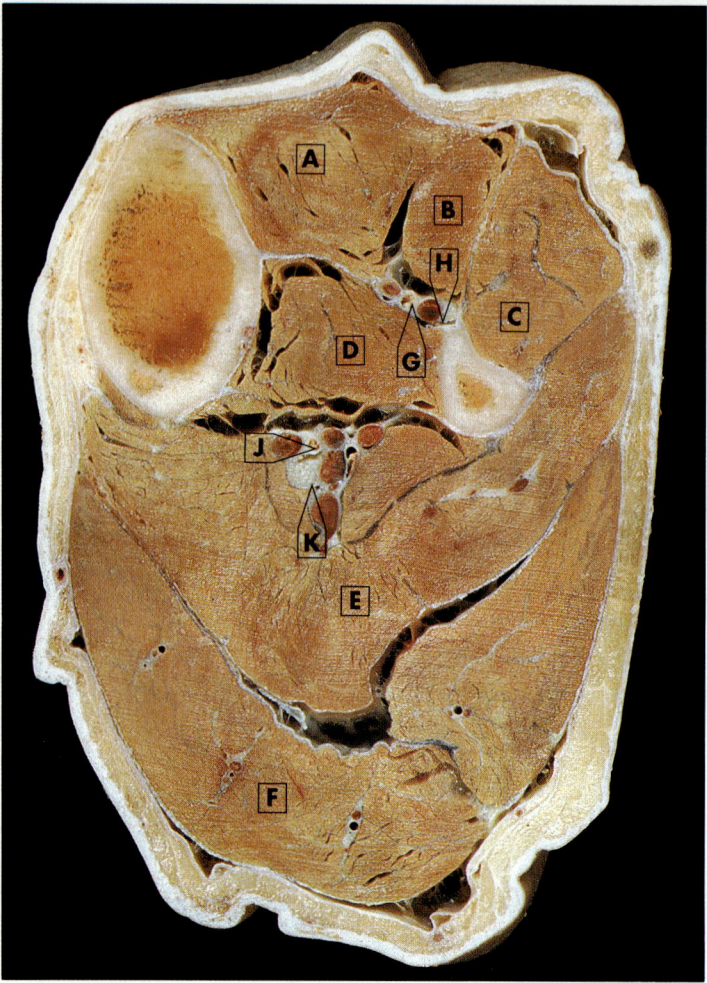

1. What are the actions of B?
2. What else lies in the compartment containing C?
3. In which compartment is E?
4. During which activities is F used?

5. How does H terminate?

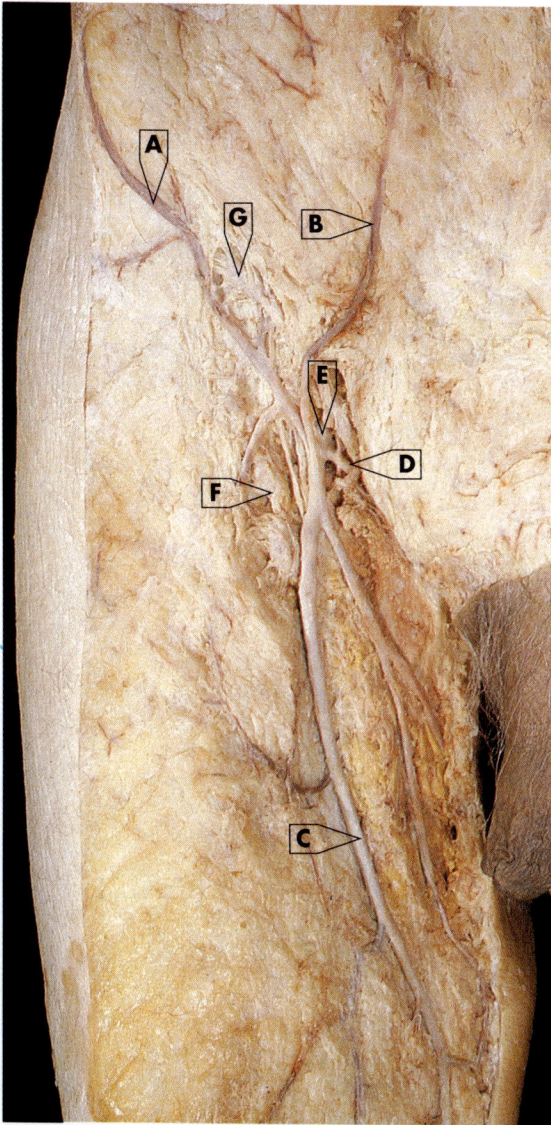

1. What is the superior limit of the region drained by B?
2. Where does C? pass deeply to enter E?
3. What is the name of the proximal continuation of E?
4. List the regions and areas drained by F and G.
5. Where do efferents from F and G drain?

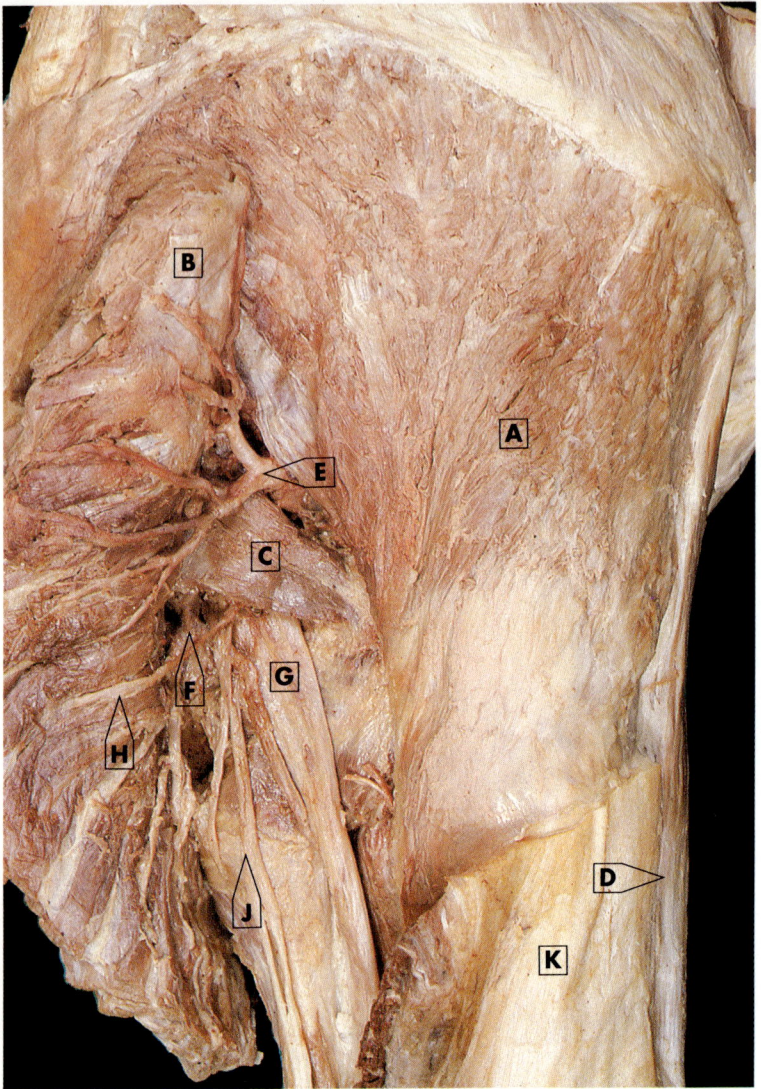

1. What are the actions of A?
2. What is the action of C?
3. What is the origin of E?

4. What is the segmental origin of G?

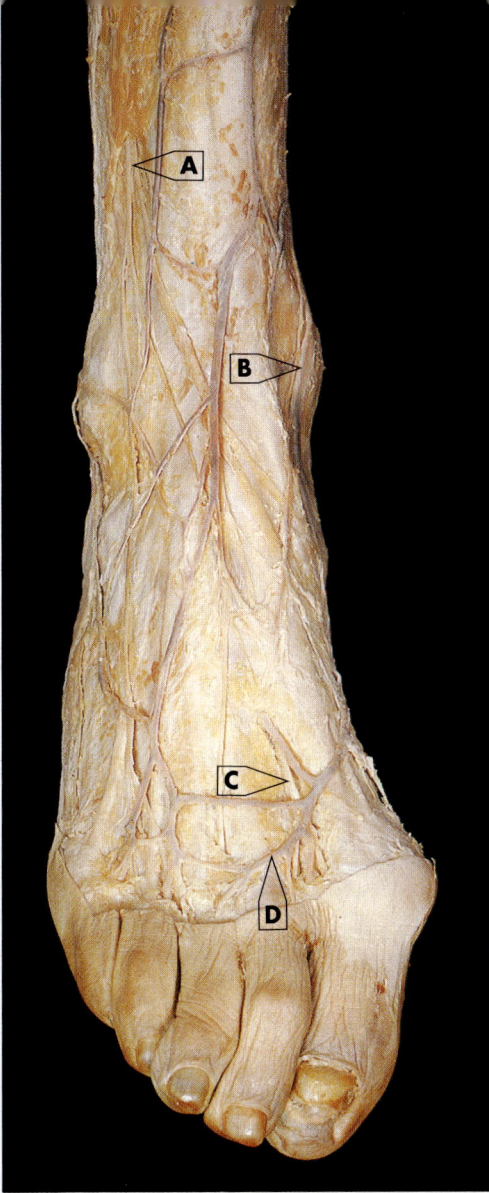

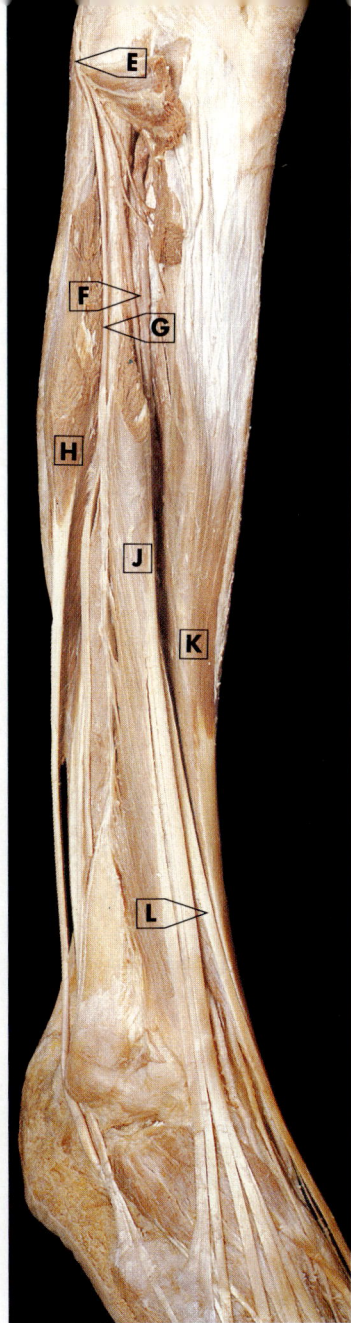

1. Where does B terminate?
2. What would be the effects of complete transsection of E?
3. Where and how does F terminate?

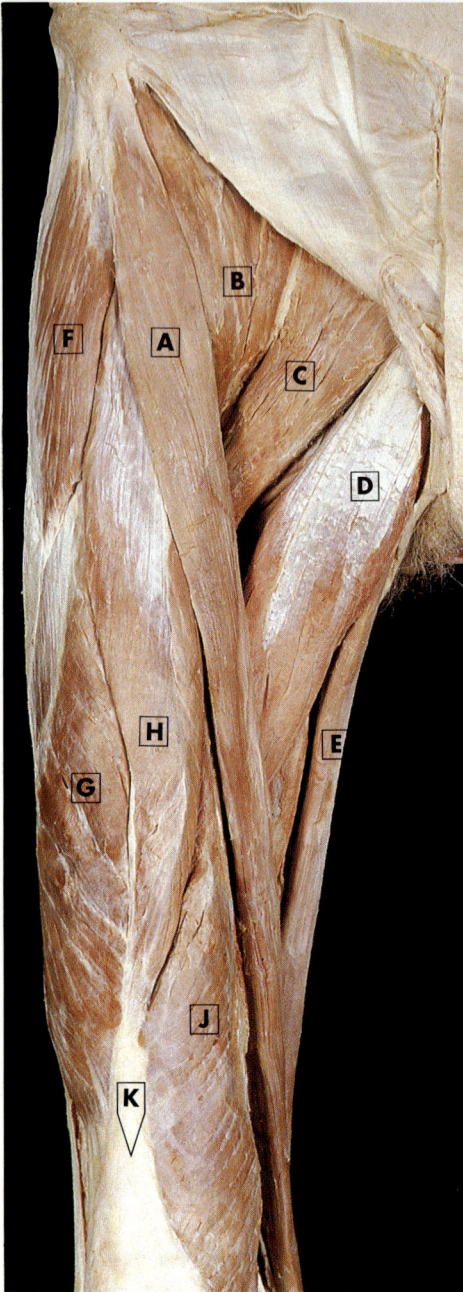

1. What are the actions of B?
2. What are the actions of D?
3. Name the muscle group to which G belongs.
4. What is the segmental origin of the nerve supplying H?
5. What is the distal attachment of K?

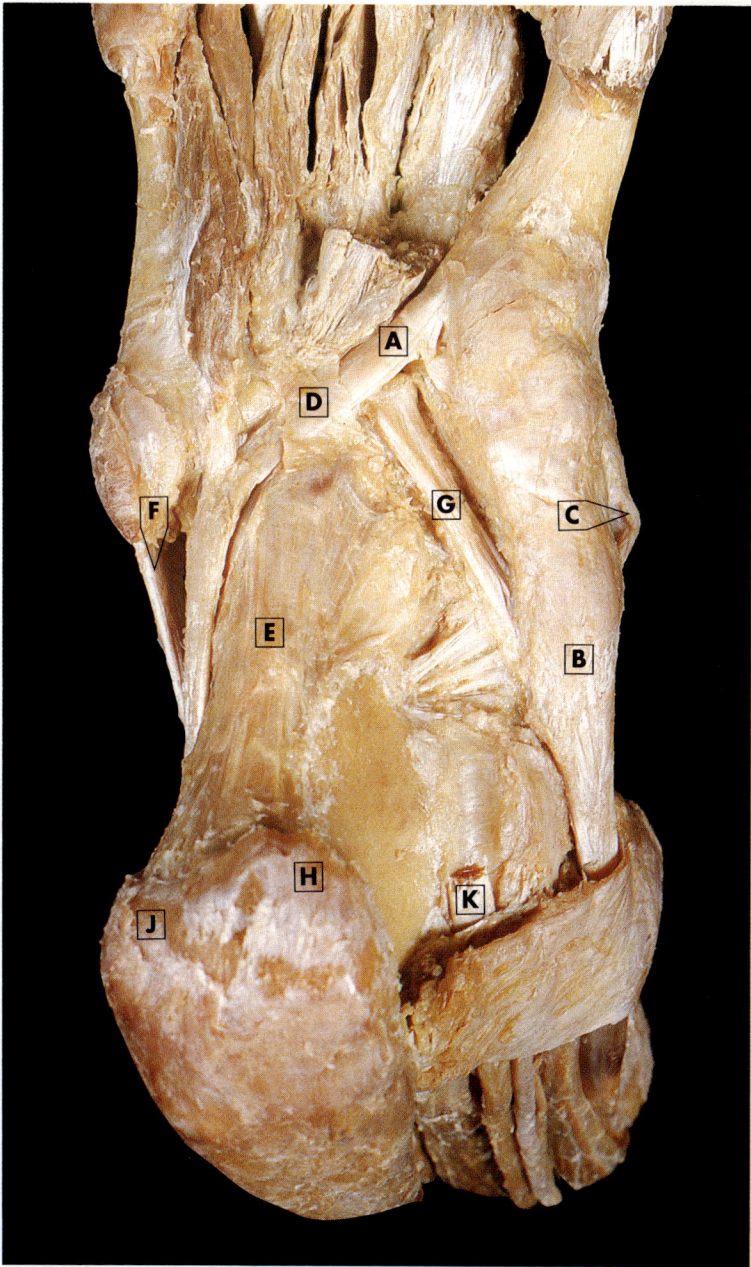

1. What are the actions of A?
2. What are the actions of B?
3. What are the functions of E?
4. What is attached at H?
5. What is the distal attachment of K?

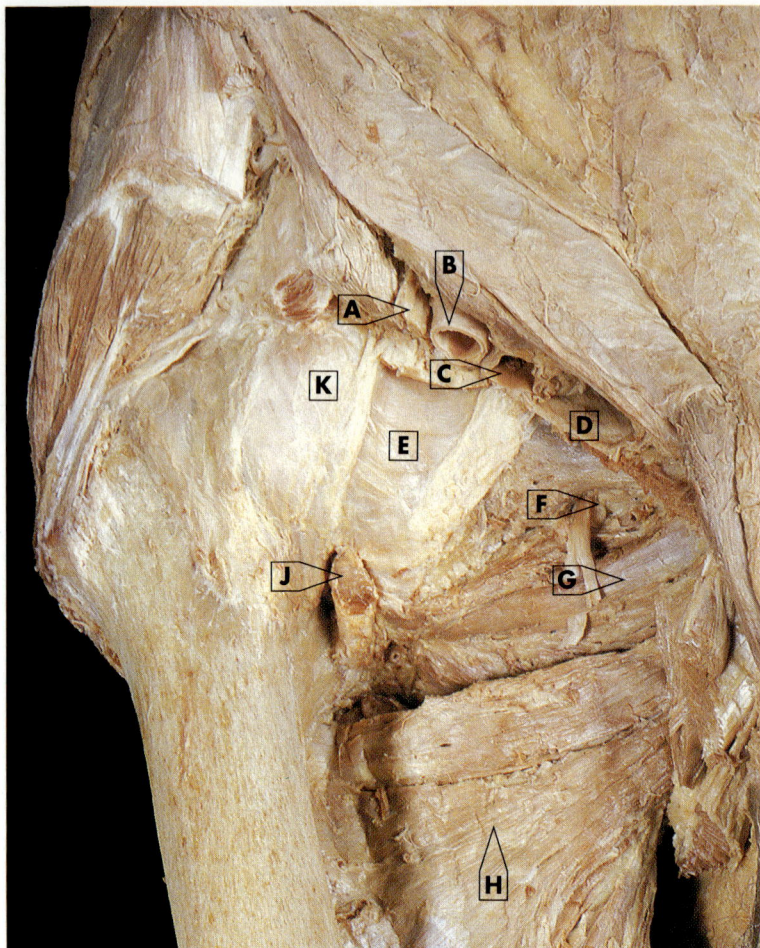

1. What is the relationship of A to the femoral sheath?
2. Is B within the femoral sheath?
3. What forms the proximal limit of D?
4. Of what does F mark the anterior limit?
5. To what is J attached?

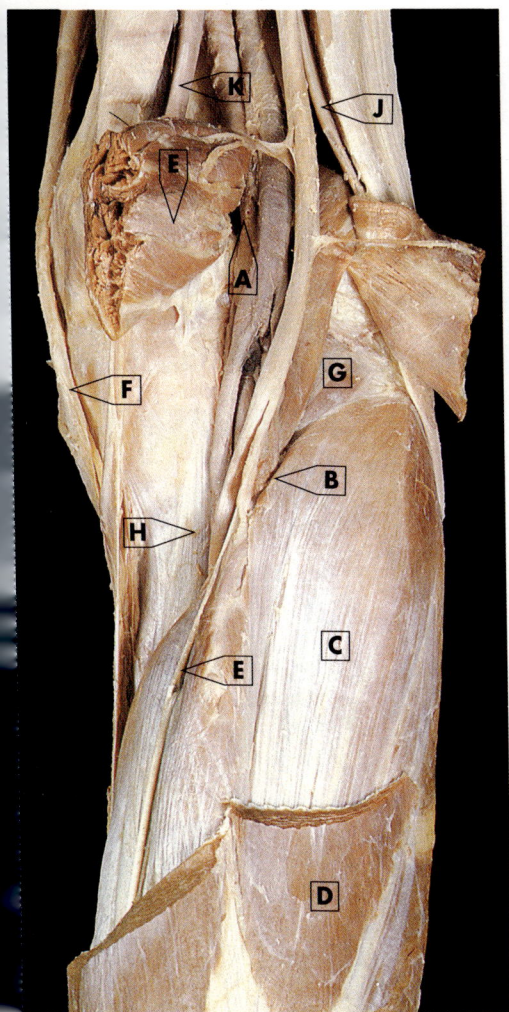

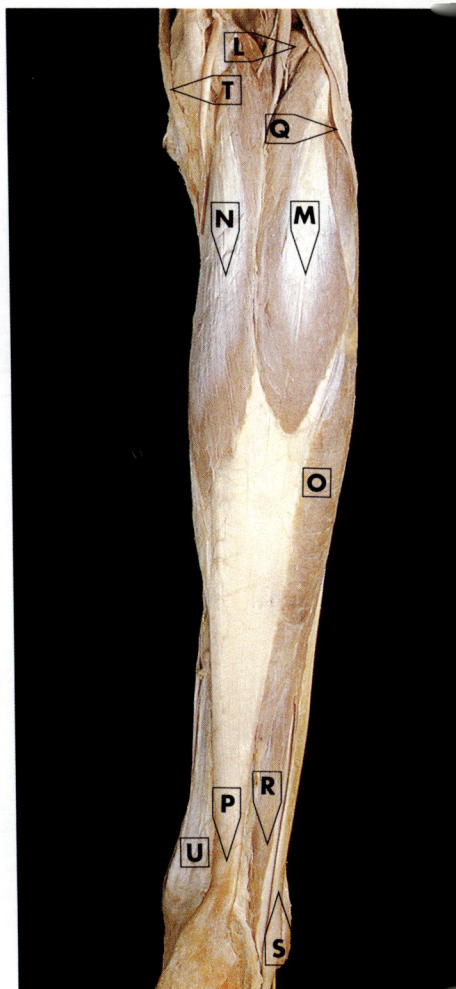

1. What is the proximal attachment of O?
2. What is the nerve supply of S?
3. What is the precise function of C?
4. What is the precise function of D compared with C?
5. What is the proximal attachment of G?

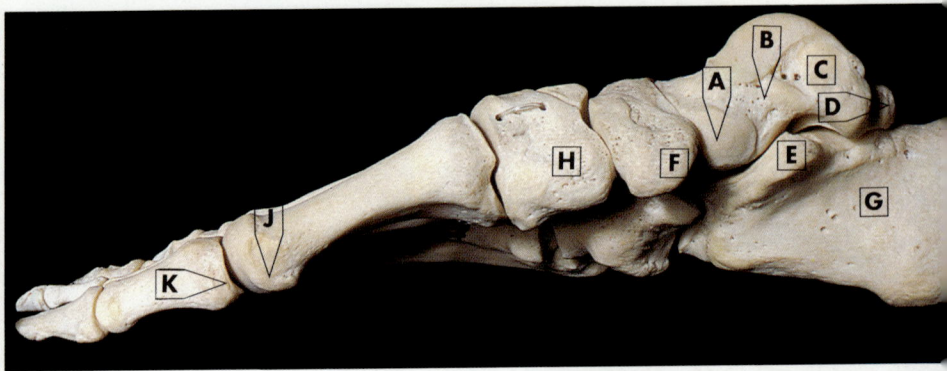

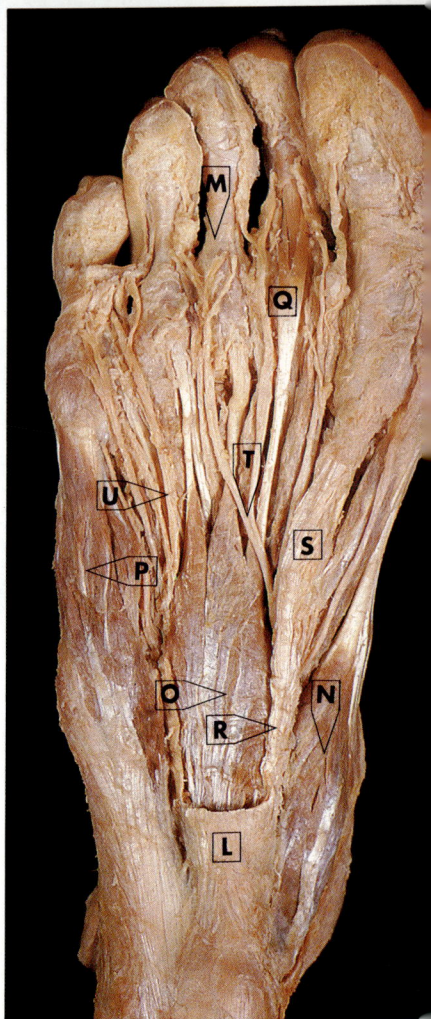

1. Which arch is illustrated?
2. What articulates with C?
3. What lies beneath E?
4. State an important function of L.
5. From where does U originate?

1. What are the actions of B?
2. What is the nerve supply of C?
3. What structure passes through E in the proximal part of the leg?
4. What are the terminal branches of G?
5. What is the nerve supply of K?

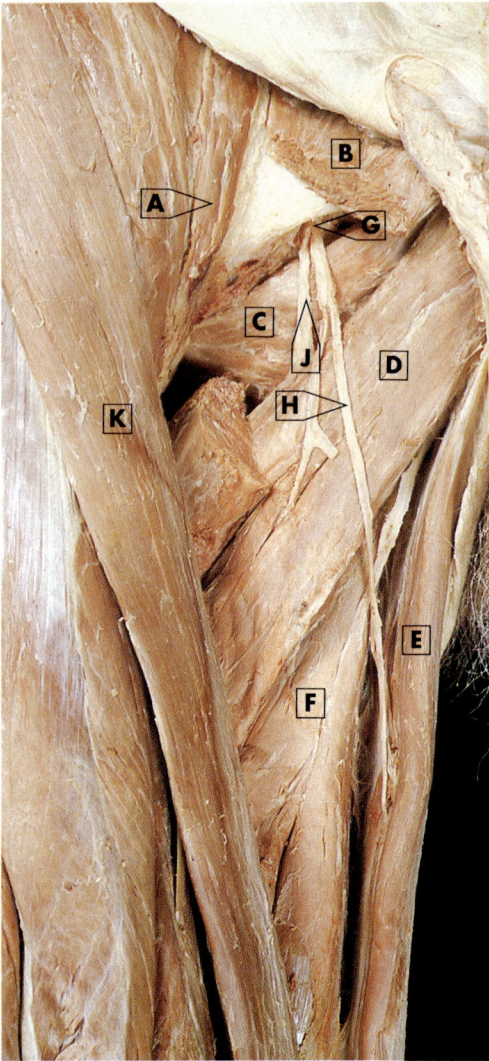

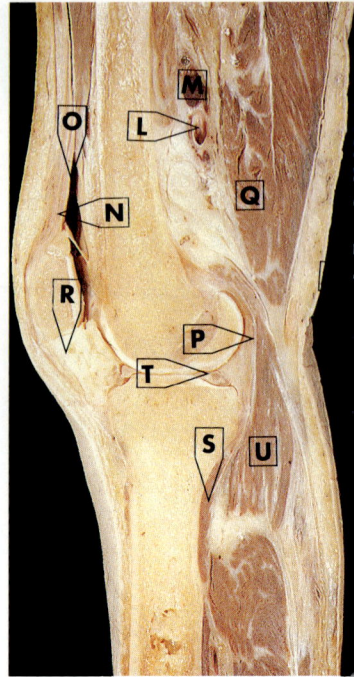

1. What is the segmental origin of G?
2. Which muscles are supplied by H?
3. What is the relationship of R to the knee joint?

4. Give the role of S.

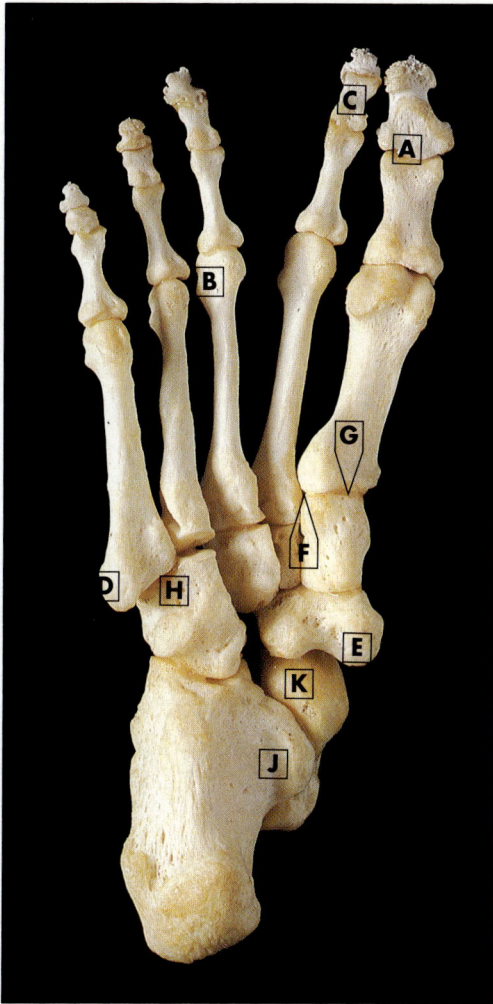

1. With what does B articulate inferiorly?
2. What attaches at E?
3. What occupies groove H?
4. What occupies groove J?

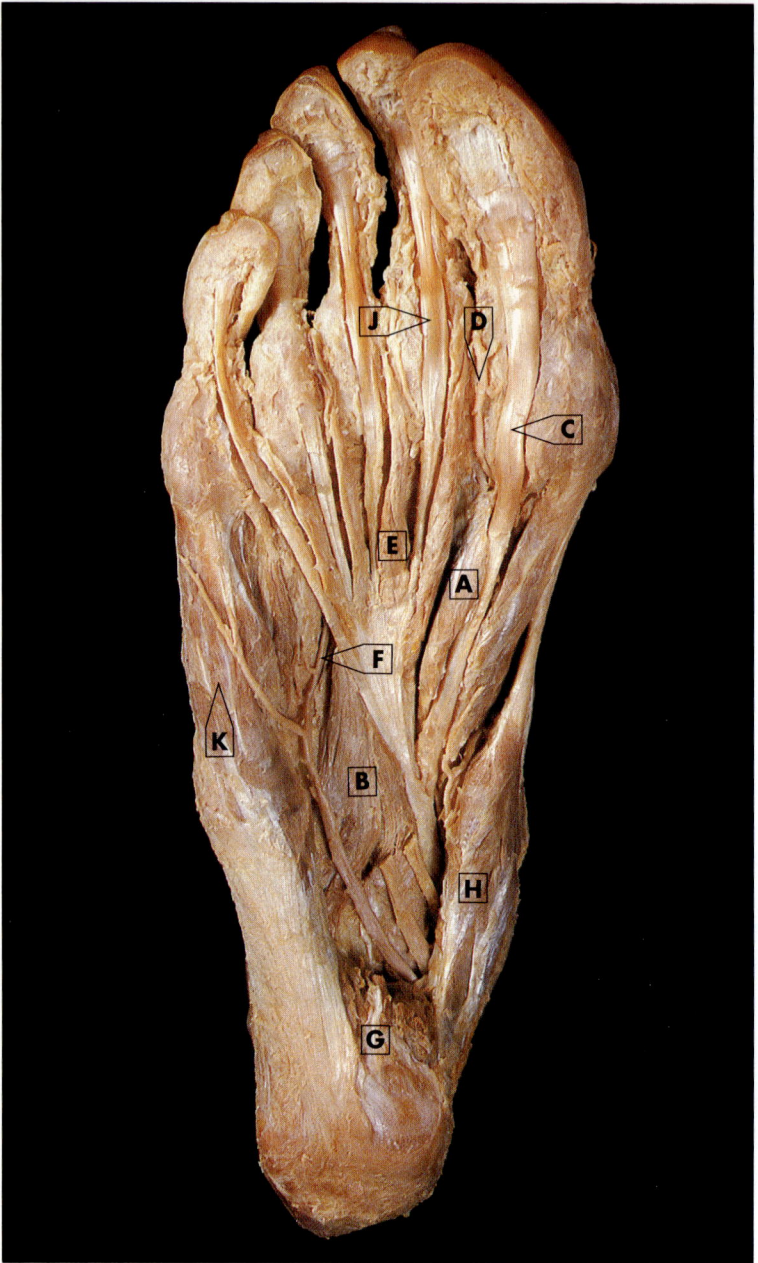

1. Give the action of A.
2. What is the role of B?
3. What is the surface landmark of the origin of F?
106 4. Which nerve supplies G?

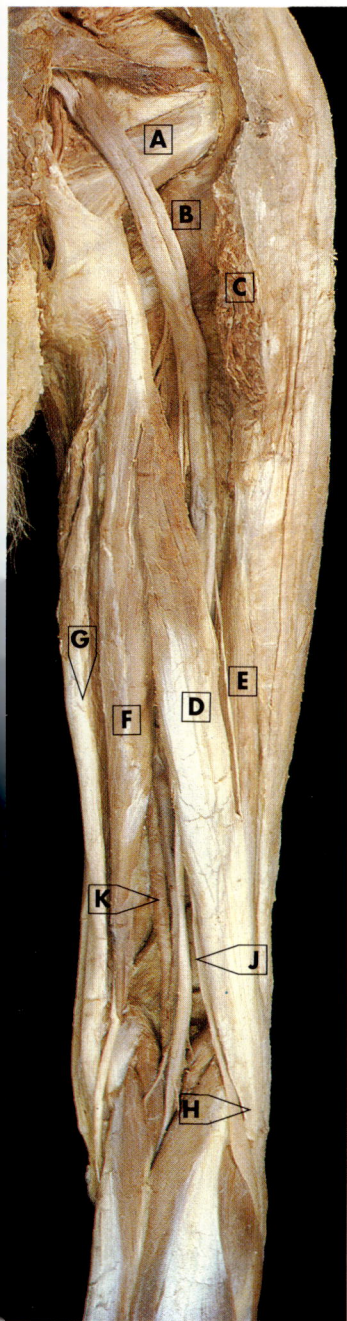

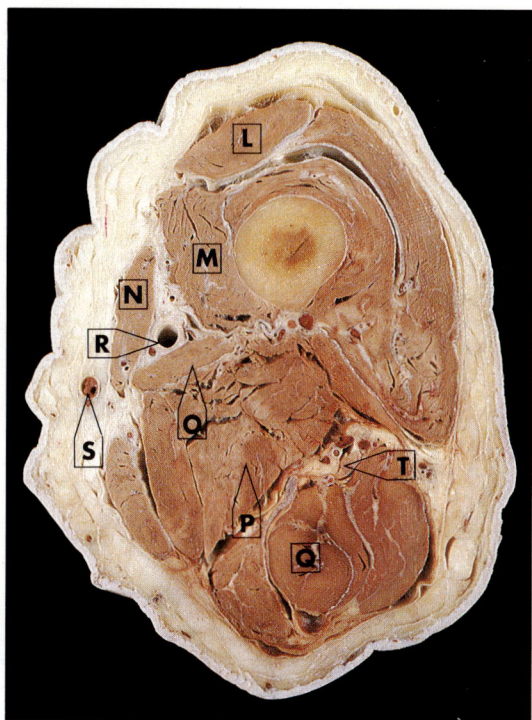

1. What is the proximal attachment of E?
2. Name the branches of H in the popliteal fossa.
3. What important function of M is not performed by other vasti?
4. Give the main actions of Q.

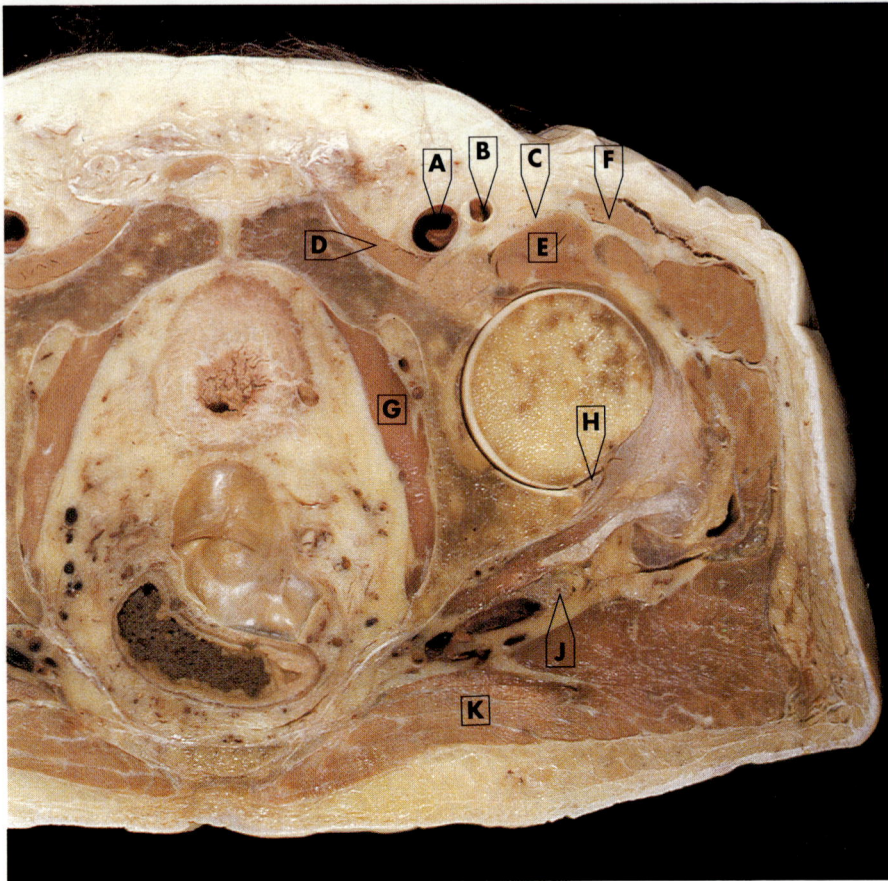

1. Which branch of C enters the leg?
2. Which nerve supplies D?
3. Give the actions of E.
108   4. What are the distal attachments of K?

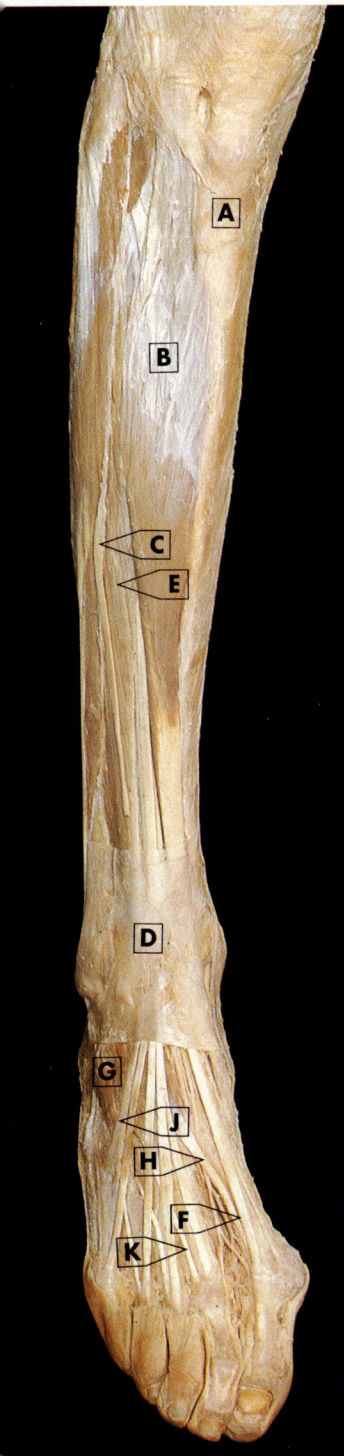

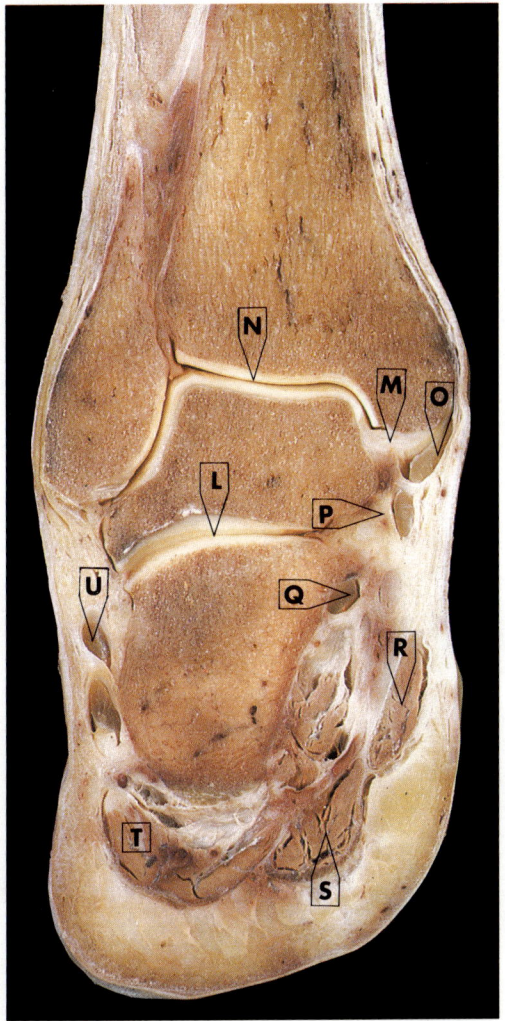

1. What are the actions of B?
2. Which nerve supplies E?
3. Classify joint L.
4. Give the function of M.

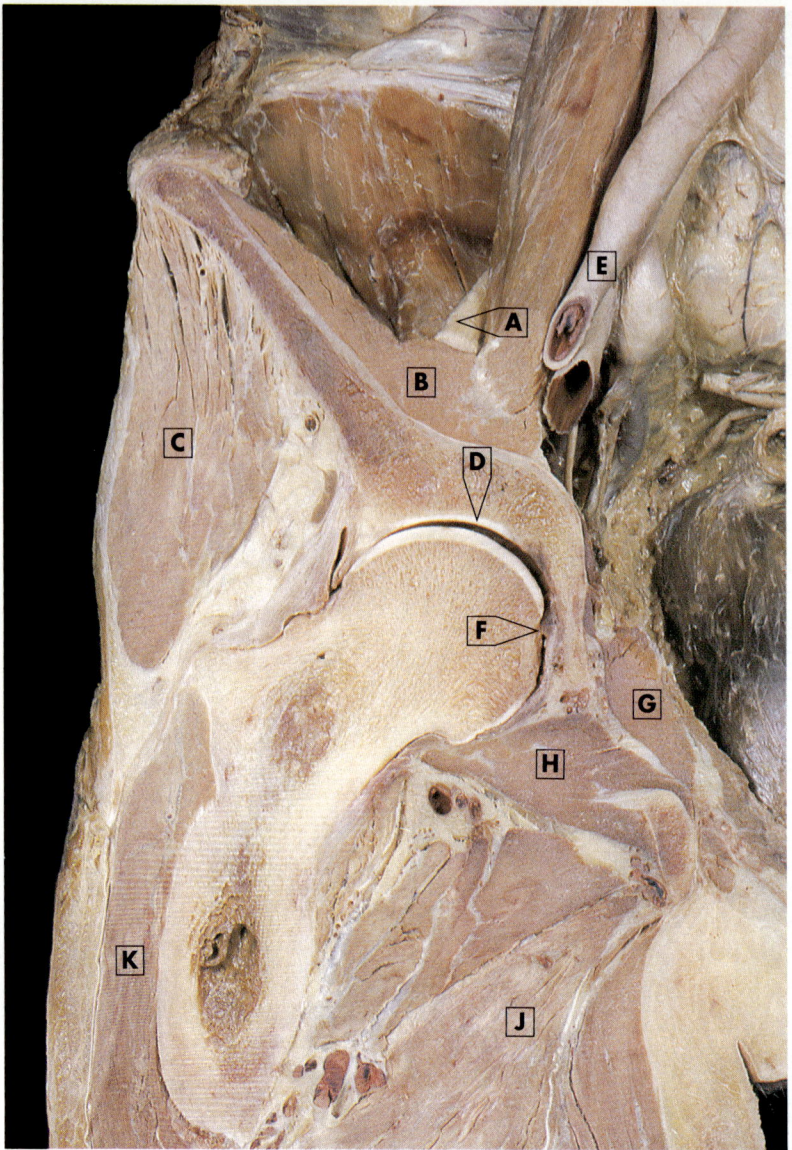

1. Which spinal cord segments are represented in A?
2. What is the distal attachment of B?
3. What is the function of F?

4. How does G reach its distal attachment?

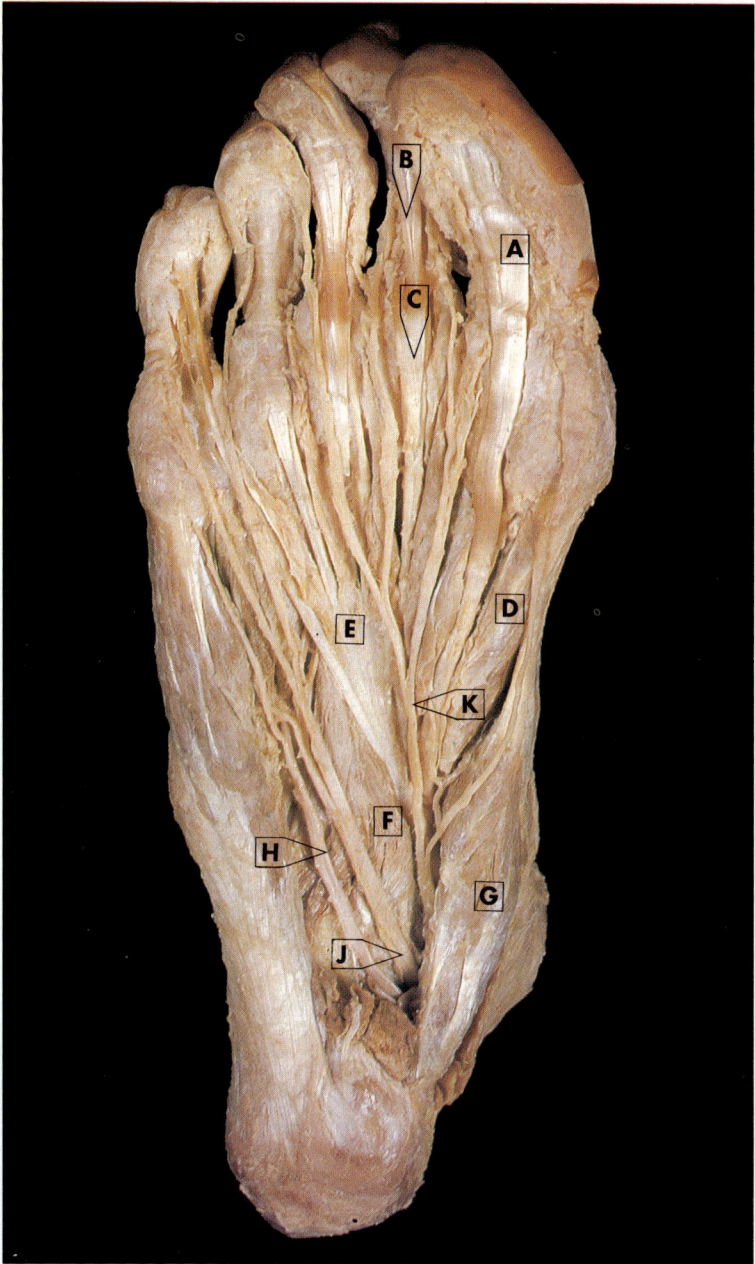

1. What is the distal attachment of B?
2. Give the actions of C.
3. How does H arise?
4. Which muscles does K supply?

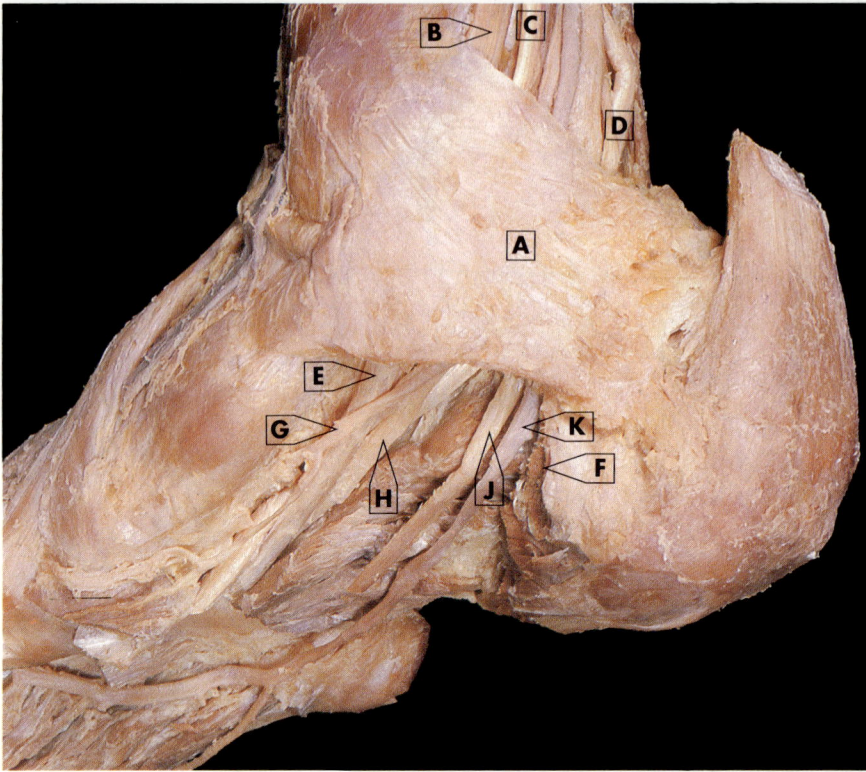

1. Name the attachments of A.
2. What are the actions of C?
3. Which nerve innervates F?

4. How does K terminate?

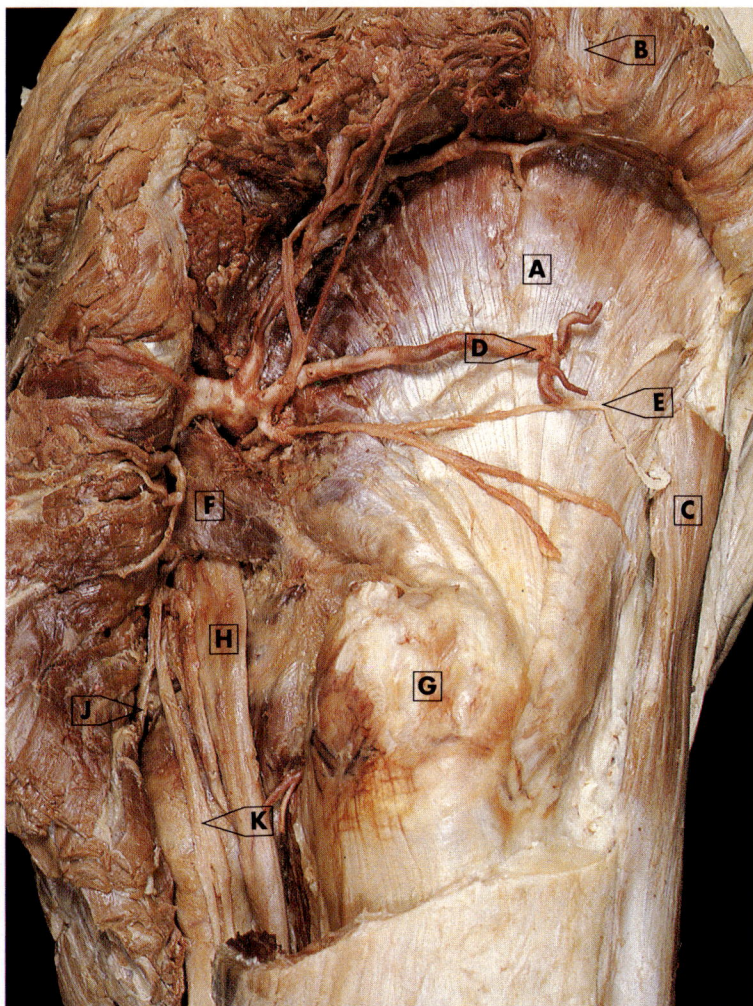

1. Give the actions of A.
2. Which nerve supplies B?
3. What is the proximal attachment of C?
4. From which segments does K arise?

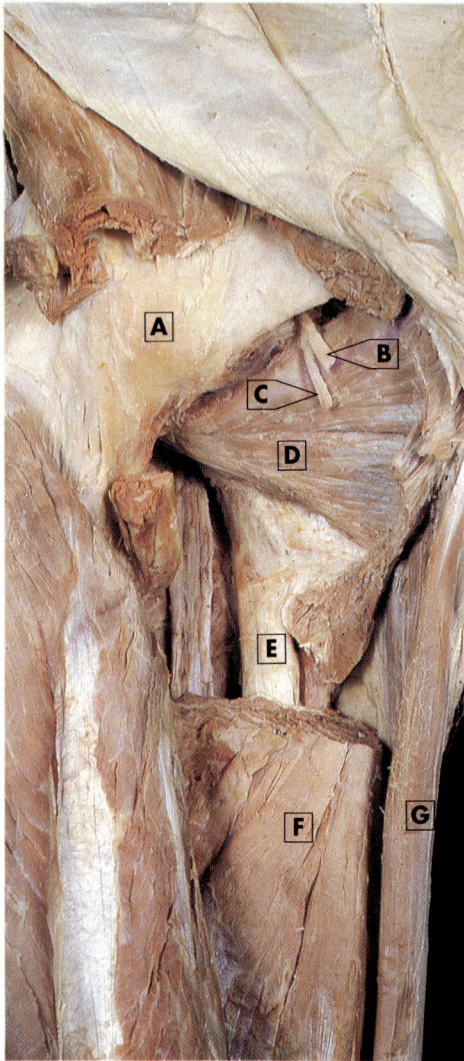

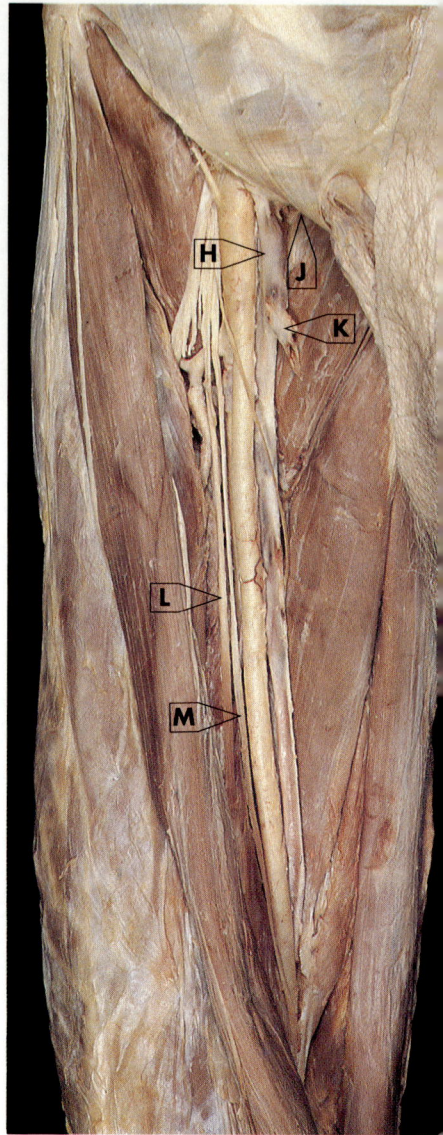

1. Apart from muscles, what does B supply?
2. What are the actions of G?
3. What is the clinical significance of J?

4. How far distally does M pass?

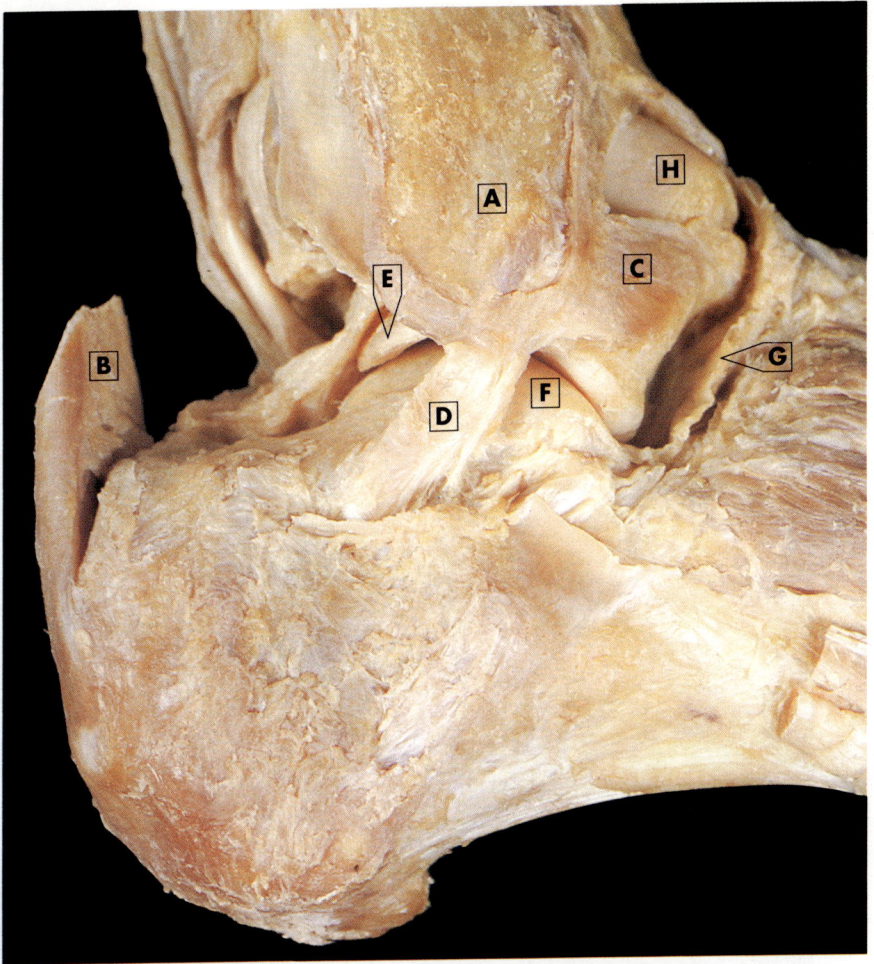

1. What superficial structures pass behind A?
2. What muscles attach to B?
3. What is the clinical importance of C?
4. With what does F articulate?

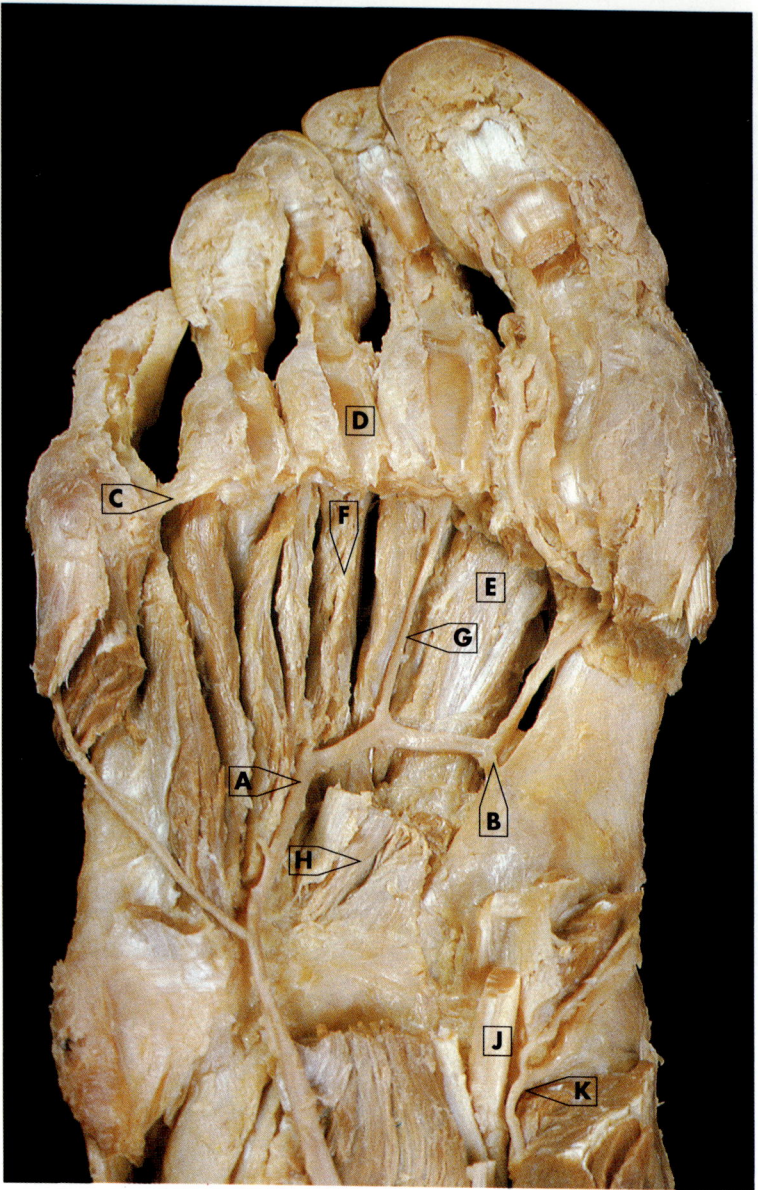

1. Where does A originate?
2. What structures pass on the dorsal aspect of C?
3. What are the actions of E?

4. What is the nerve supply of H?

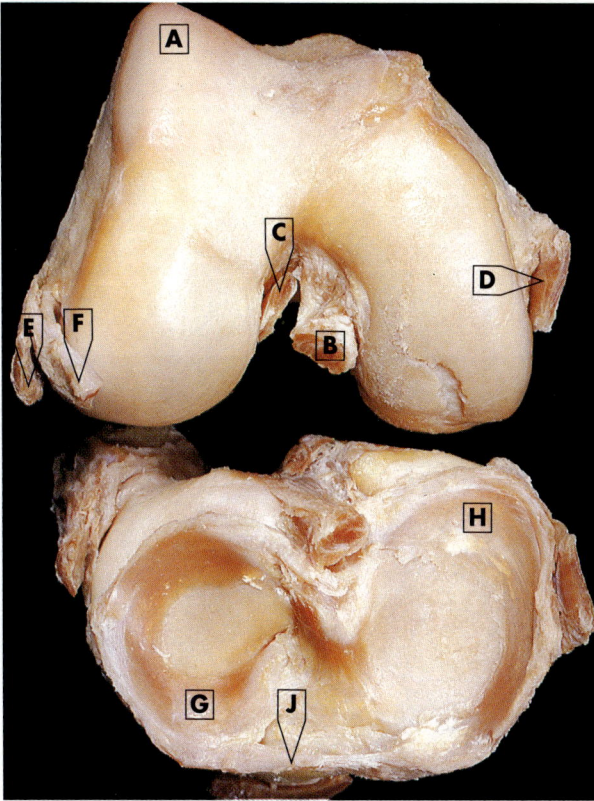

1. Why is A so prominent?
2. What are the functions of C?
3. What is the proximal attachment of E?
4. How are G and H attached to the tibia?

1. What movements occur at A?

2. What movements occur at B?

3. What is the nerve supply of F?

4. What is the function of G?

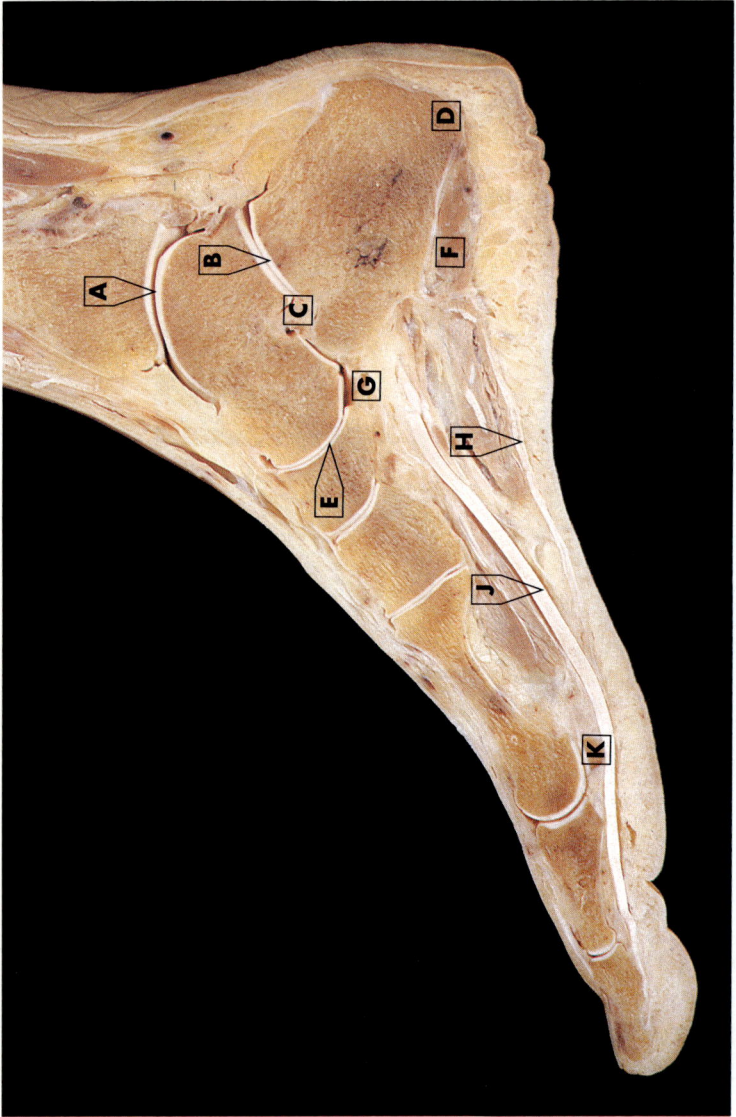

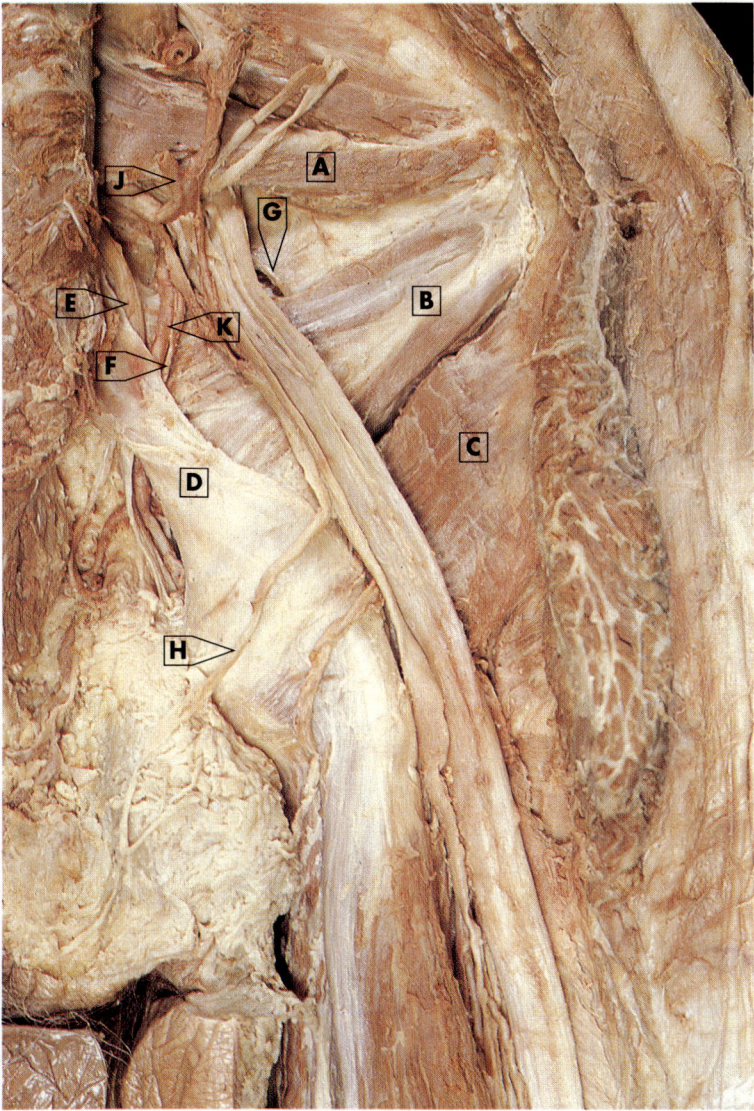

1. How does A enter the gluteal compartment?
2. What is the action of C?
3. What are the main branches of E?
4. What is the origin of K?

119

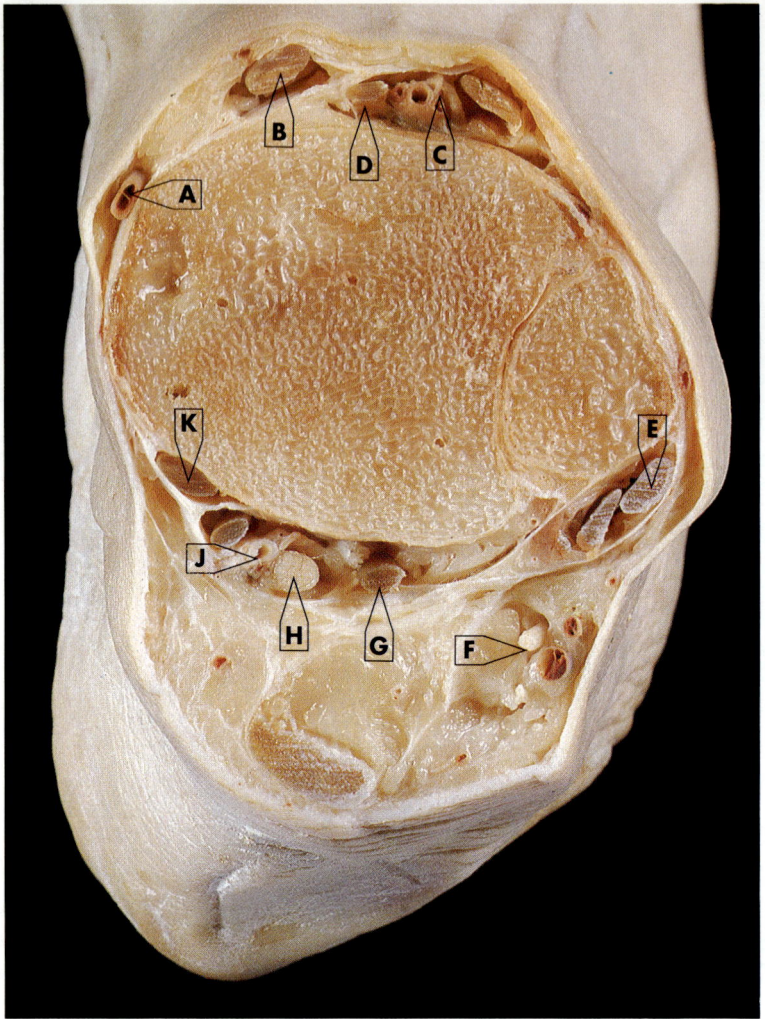

1. Where does A originate?
2. Where does F terminate?
3. What are the terminal branches of H?

4. Where does J originate?

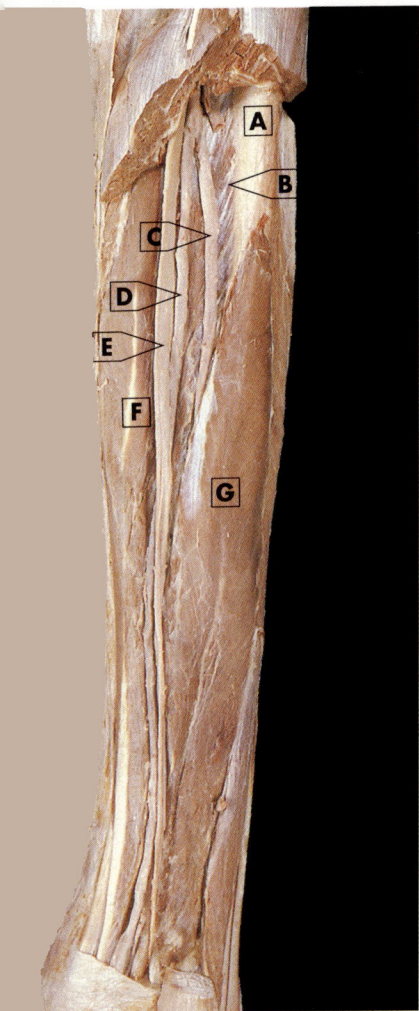

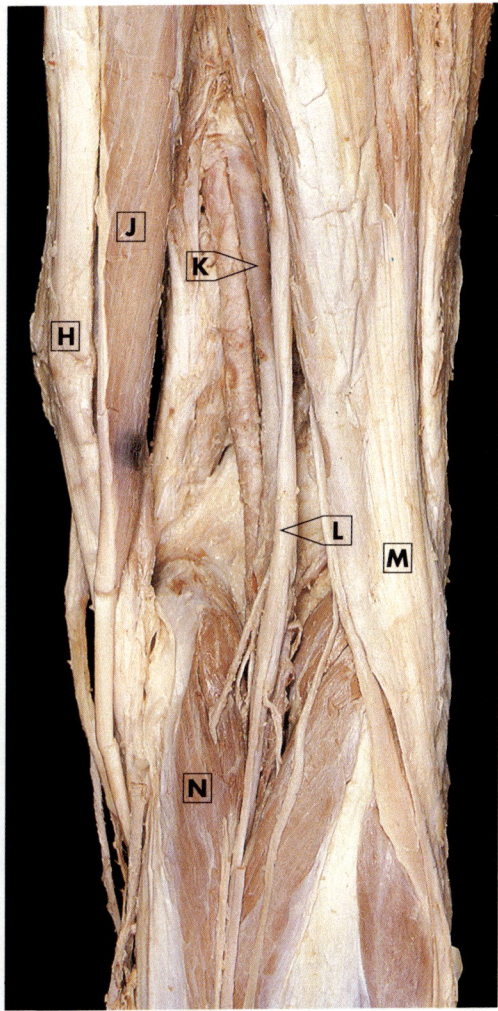

1. What neurovascular structure is closely related to A?
2. What are the actions of B?
3. What is the nerve supply of J?
4. What is the distal attachment of M?
5. What is the distal attachment of N?

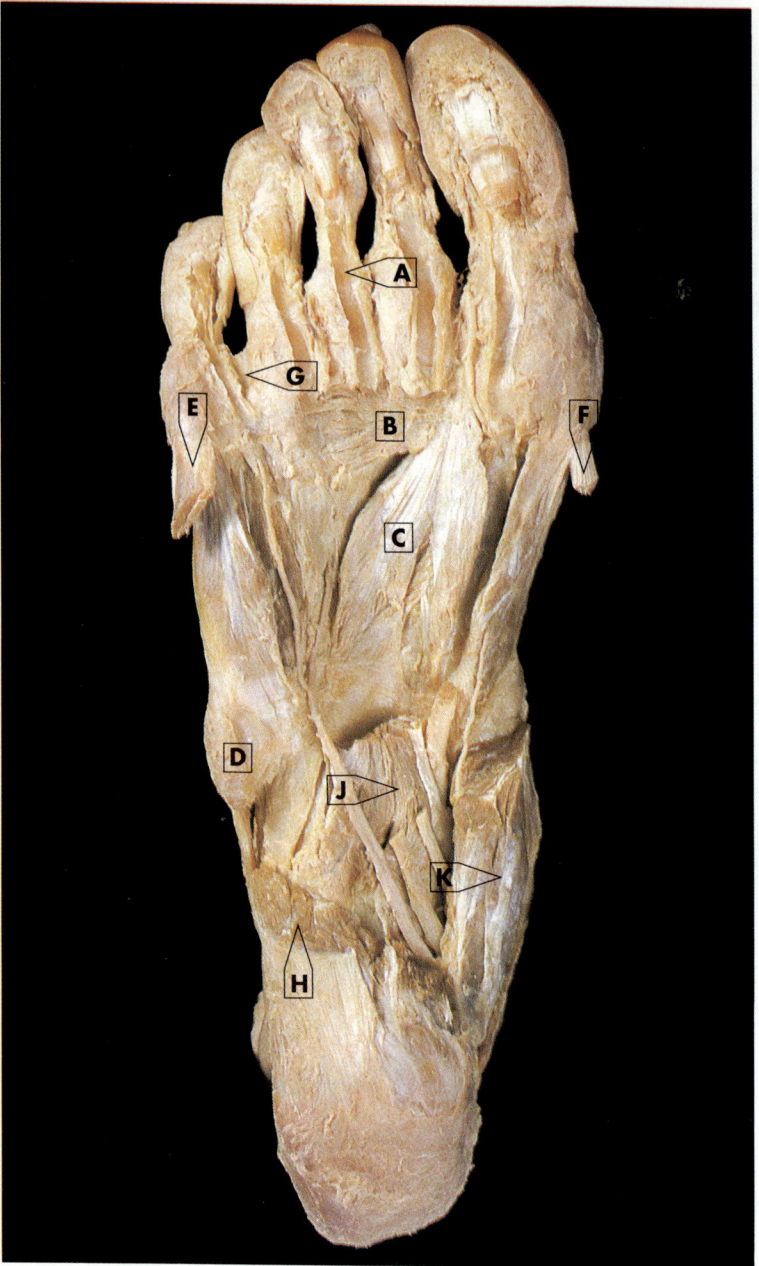

1. What is the nerve supply of B?
2. What attaches to D?
3. What lies immediately superficial to H?

4. Which dermatome overlies K?

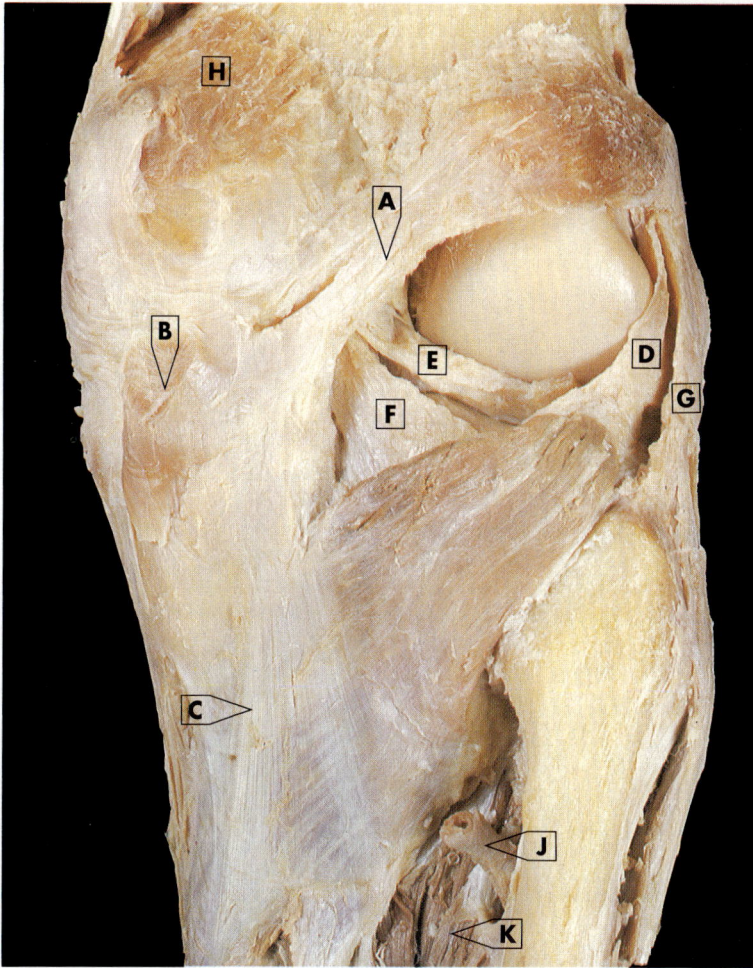

1. What structures lie in contact with C?
2. What is the function of D?
3. What is the function of F?
4. What is the termination of J?

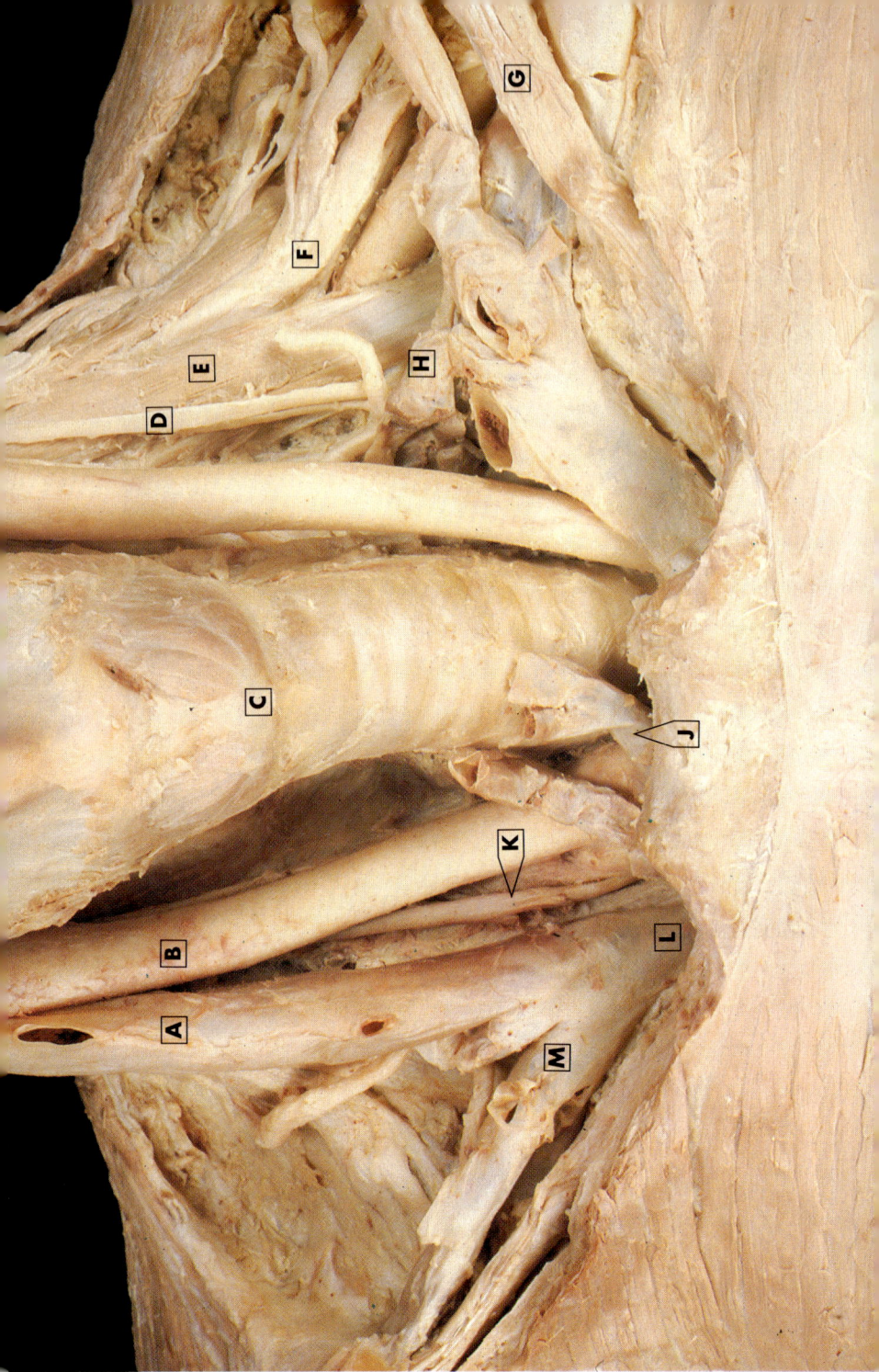

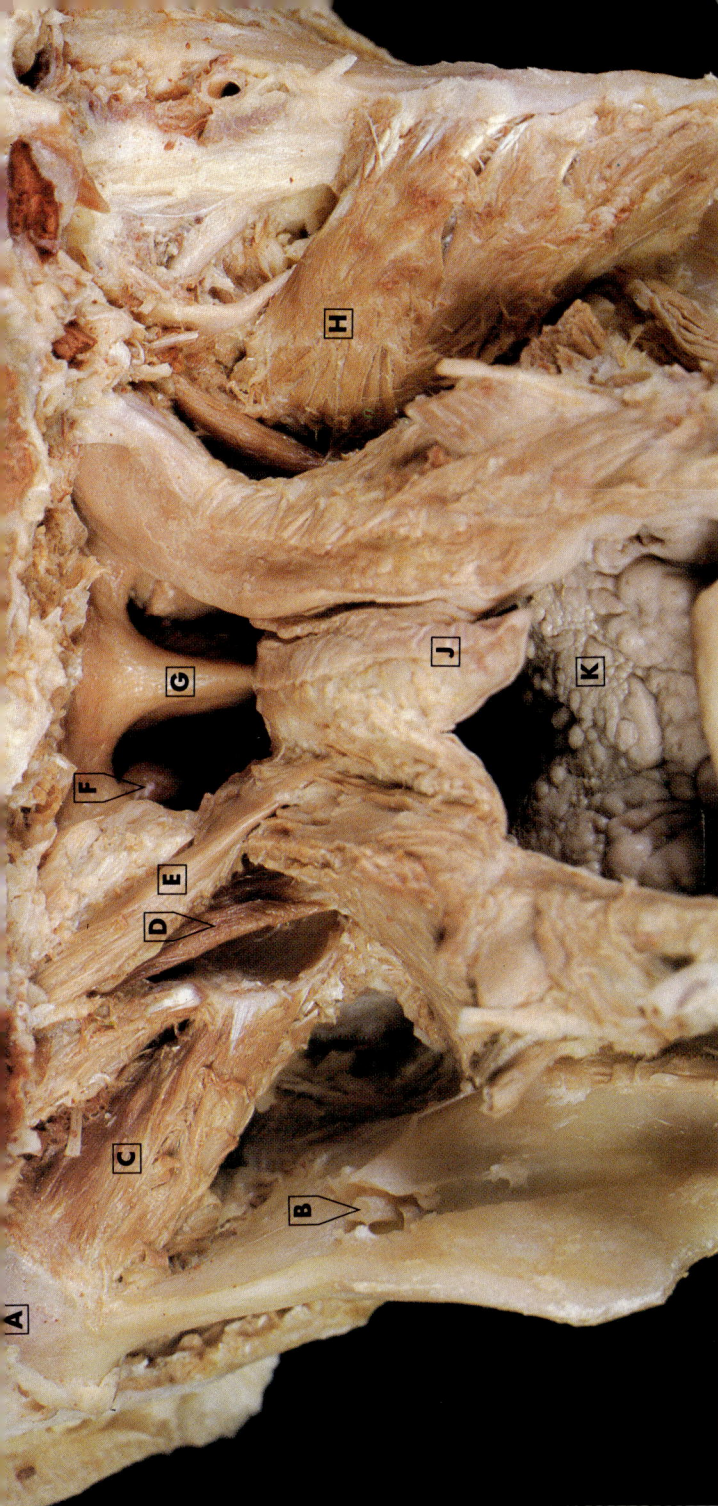

1. What is the motor nerve supply to C?
2. What is the motor nerve supply to E?
3. Which bone lies in G?
4. What is the action of H?
5. What is the sensory nerve supply to K?

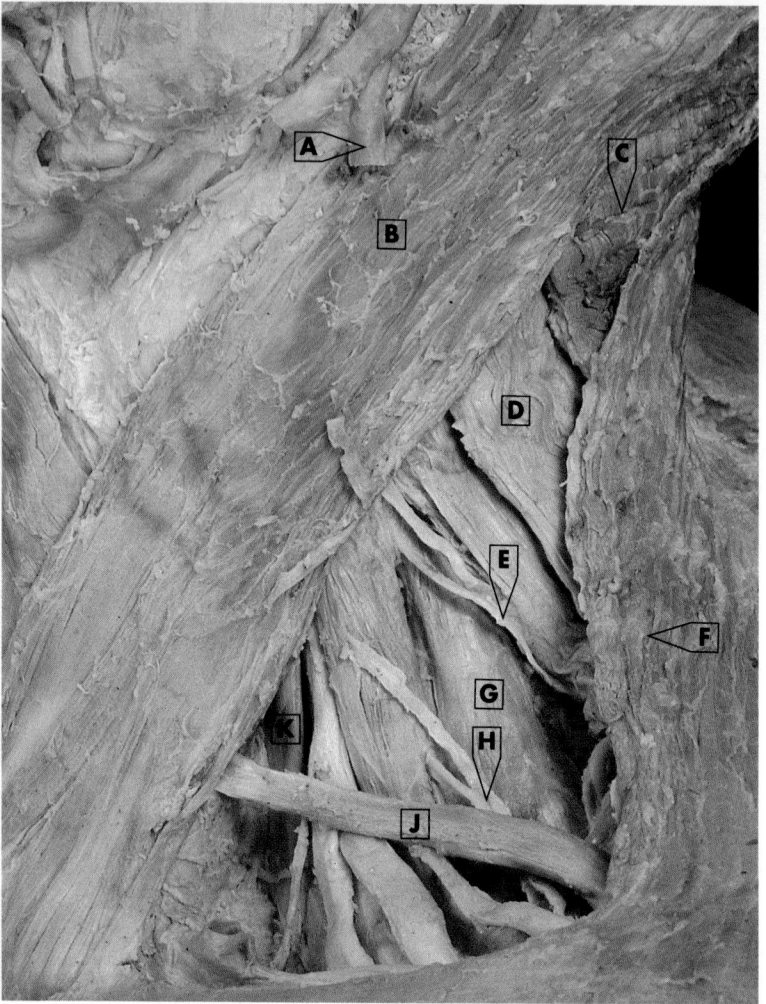

1. Which layer of fascia immediately overlays C, D, G & K?
2. What is the motor supply to B?
3. Which cranial nerve, not present here, crosses between B & F?

4. What is the motor supply to J?

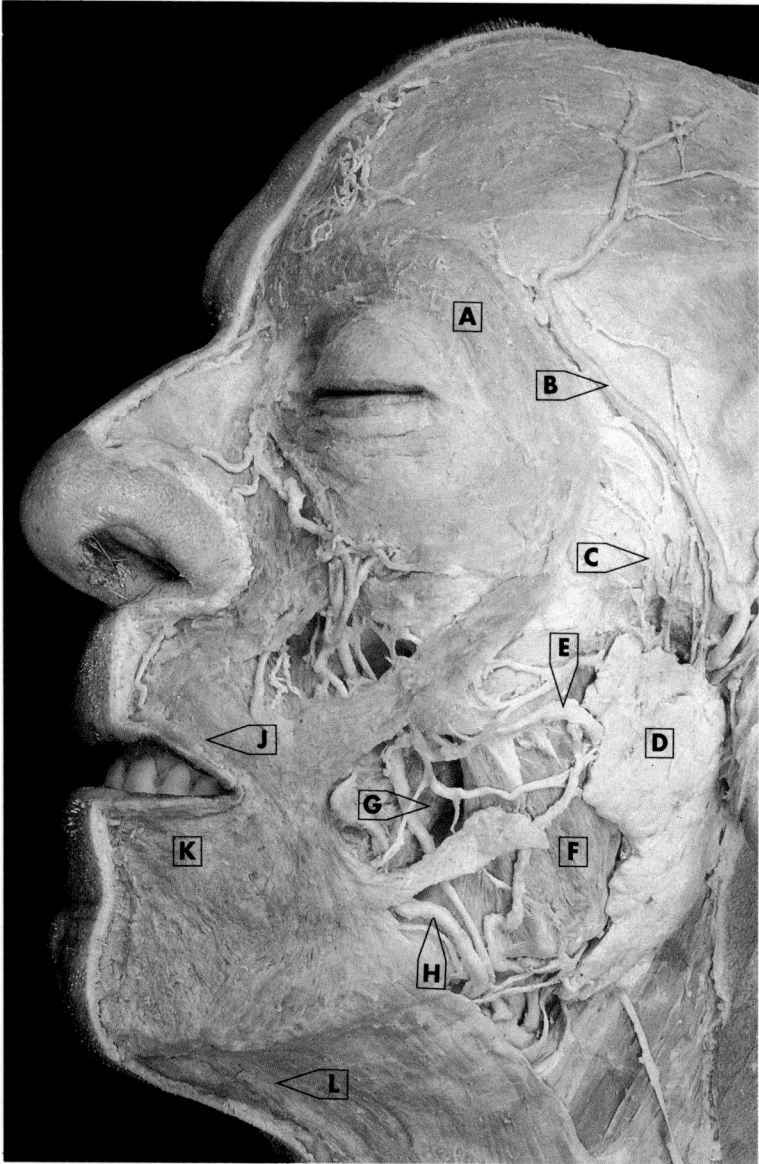

1. What is the motor supply to A?
2. What is the origin of B?
3. What is the sensory nerve supply at A?
4. What is the secretomotor supply to D?
5. What is the motor supply to F?

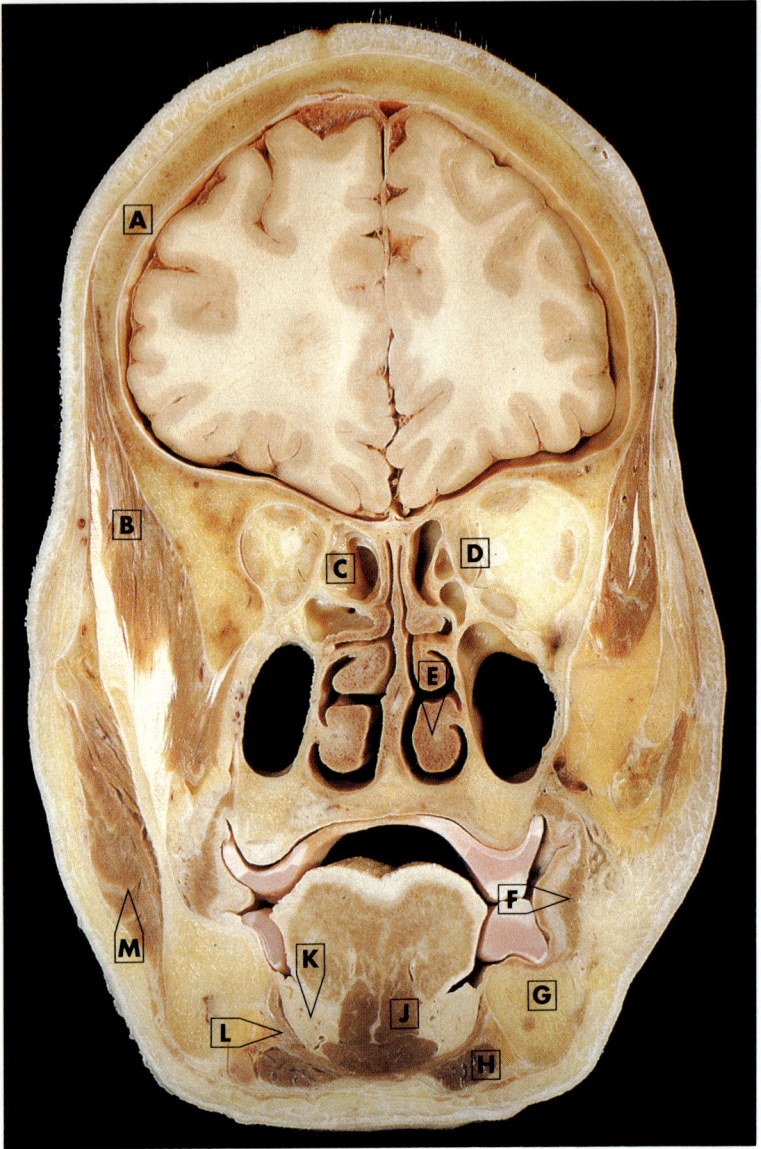

1. Where does C drain?
2. Which bone lies in E?
3. What is the motor nerve supply to F?
4. Which nerve lies within G?

5. Where does K drain?

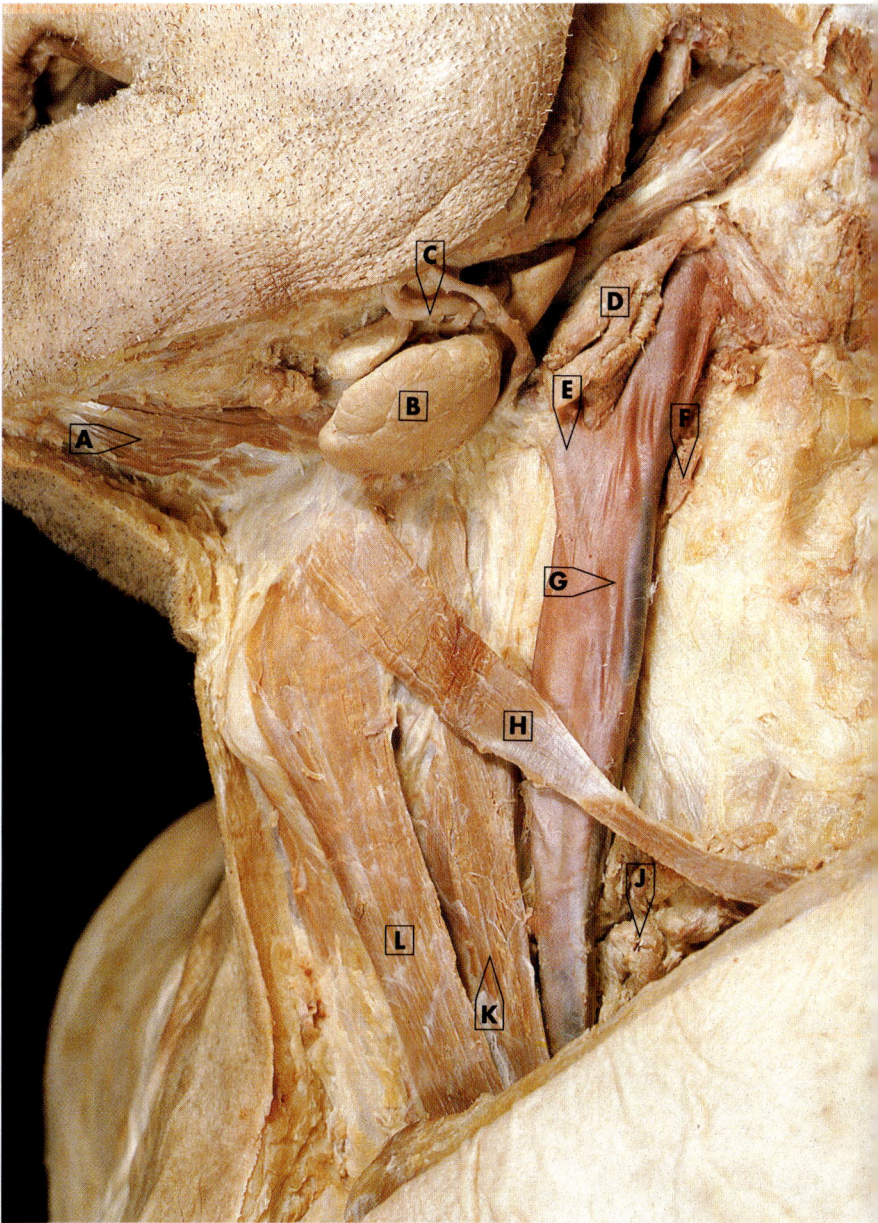

1. What is the motor supply to A?
2. Where does lymph from F drain?
3. Where does G drain?

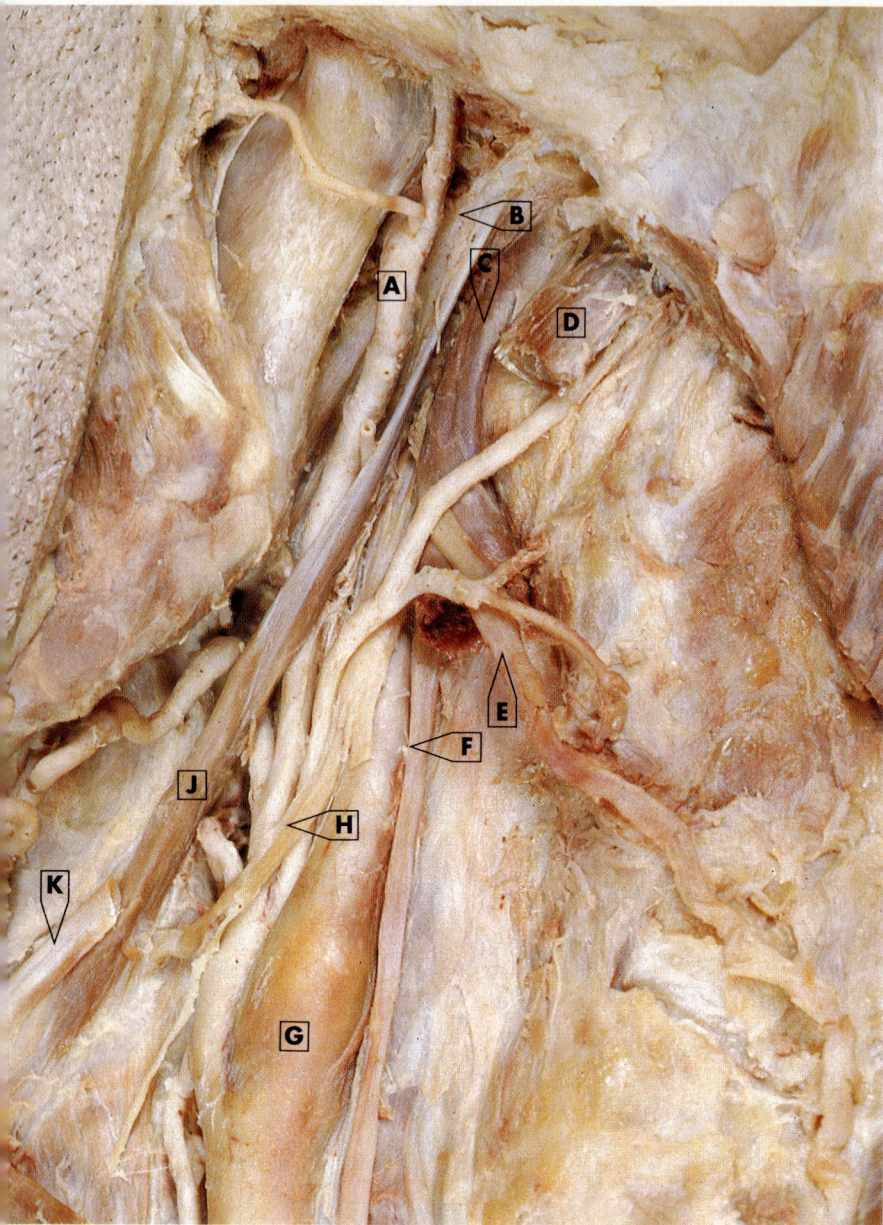

1. What is the termination of A?
2. What muscles are attached to B?
3. What is the motor supply of J?

4. What is the bony attachment of K?

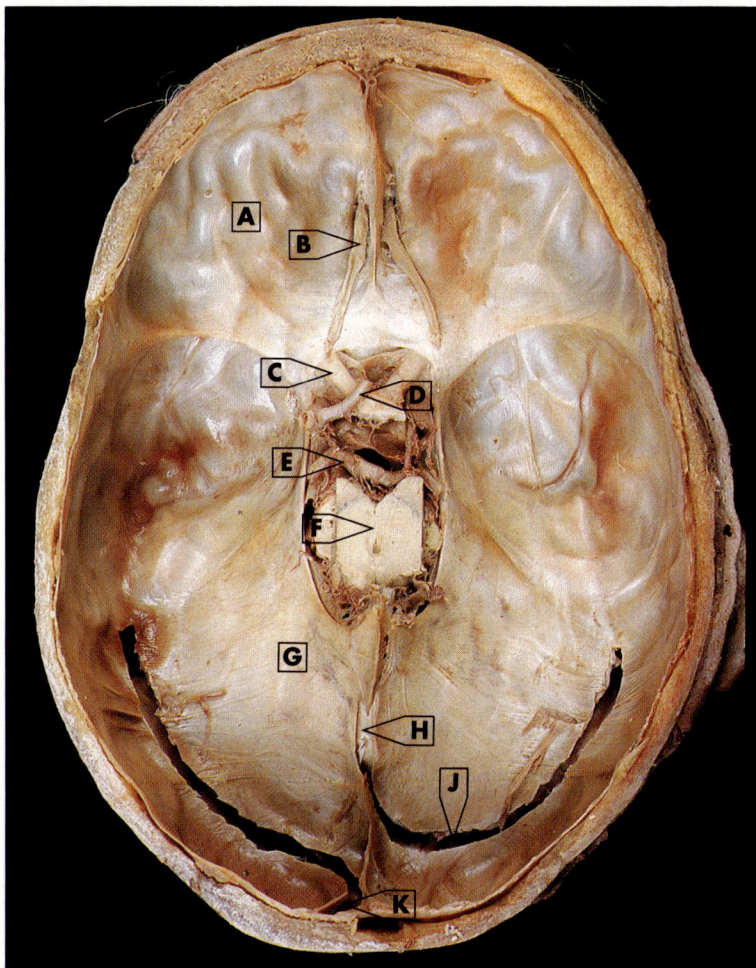

1. Which part of the brain occupies A?
2. Through which opening does C leave the cranial cavity?
3. What is the origin of D?
4. What lies below G?
5. Where does J drain?

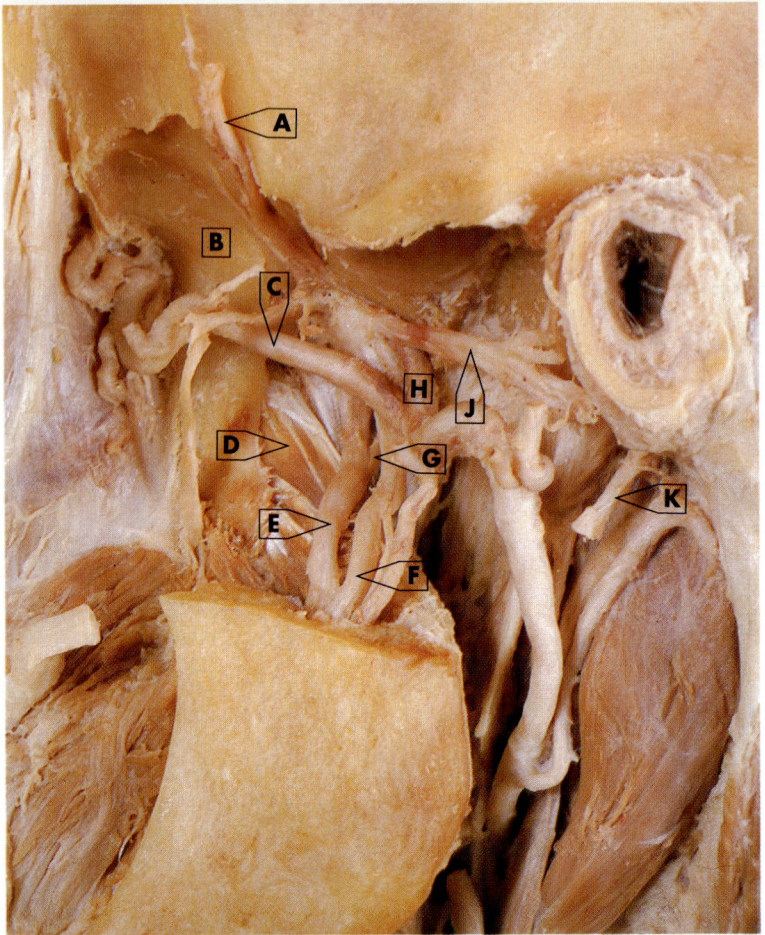

1. What is the distribution of A?
2. What is the action of D?
3. What is the distribution of F?
4. What kind of fibres are in G?

5. What is the distribution of H?

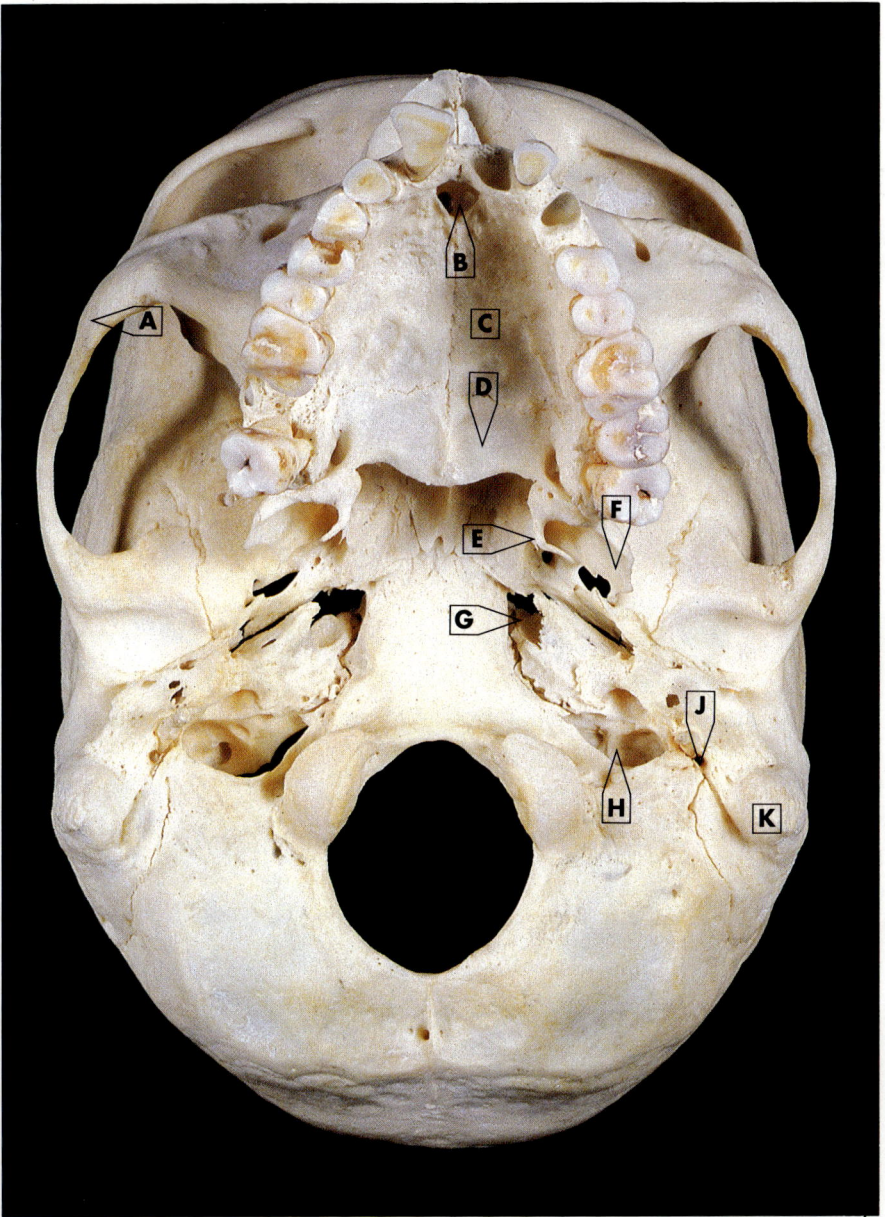

1. What passes through B?
2. What hooks round E?
3. What attaches to the lateral surface of F?
4. What emerges from the inferior surface of G?
5. What emerges from J?

133

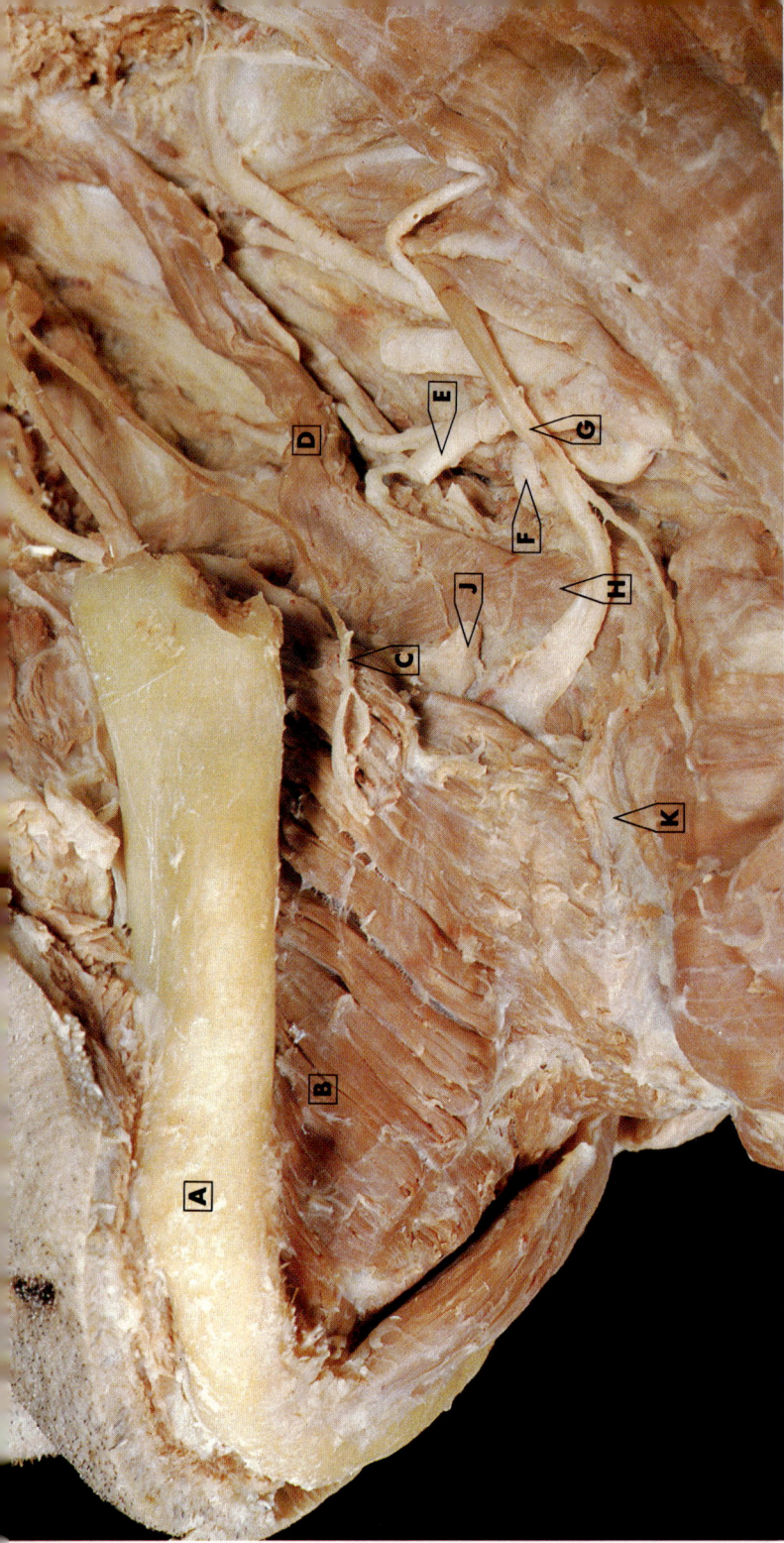

1. Which cranial nerve supplies B?
2. What is the motor supply to D?
3. Where does E cross the mandible?
4. What is the action of H?

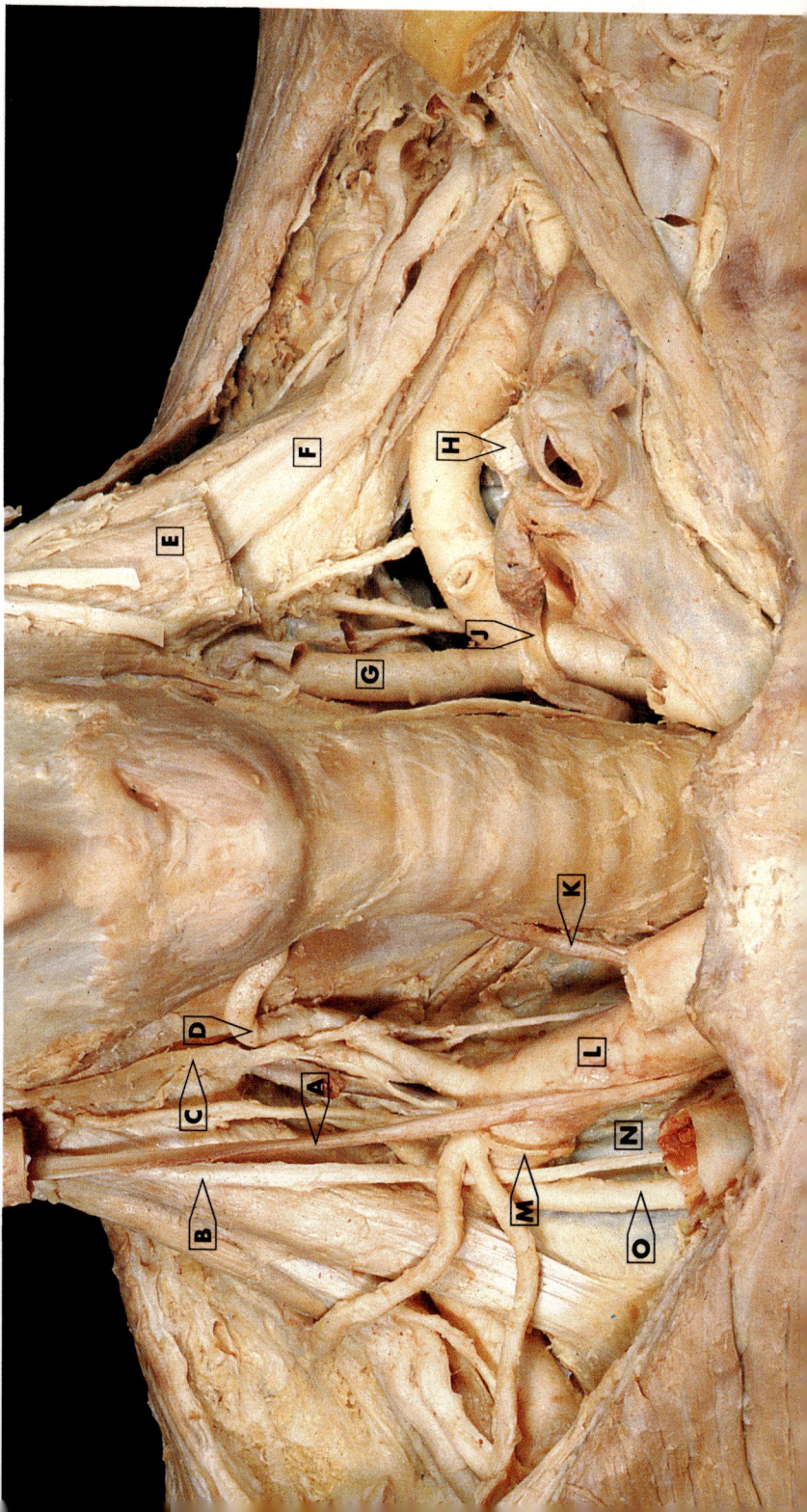

1. What lies deep to N?

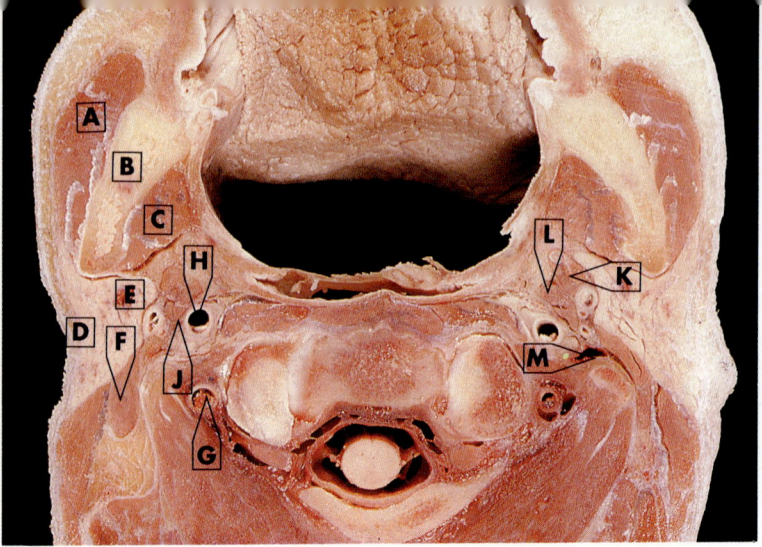

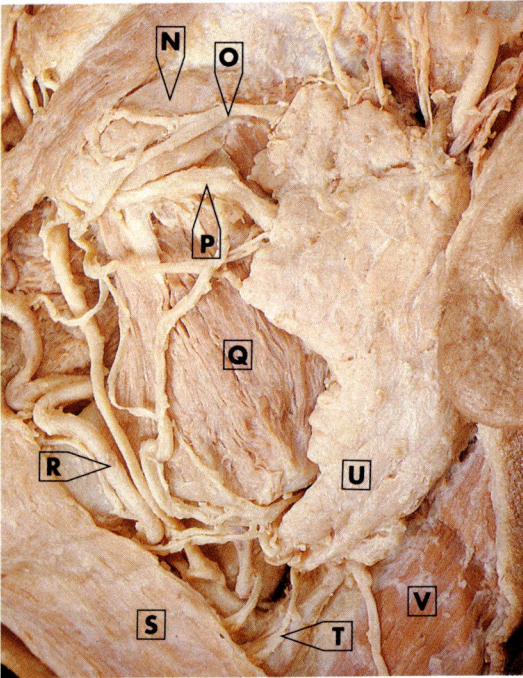

1. What is the action of A?
2. Where does D drain?
3. What is the termination of G?
4. What is the motor supply to J?
5. What is the motor supply to K?
6. What is the motor supply to L?
7. What is the termination of M?

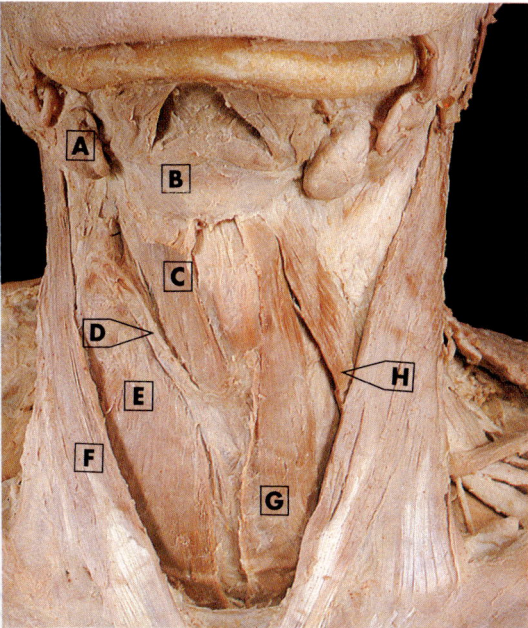

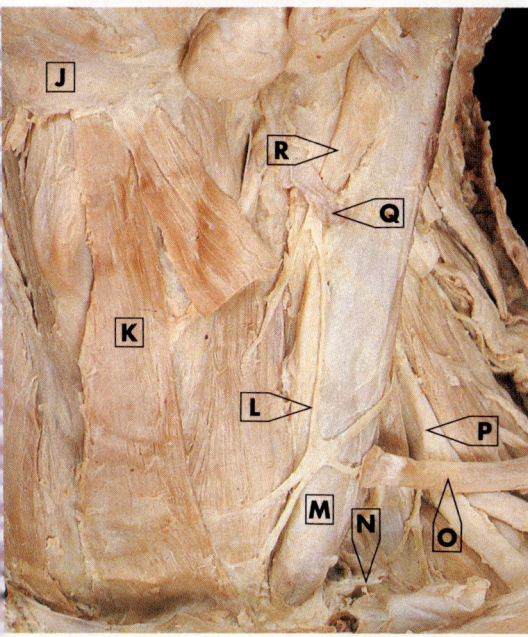

1. What is the motor supply to C?
2. What is the action of F?
3. What is the posterior attachment of H?
4. What is the segmental origin of L?
5. What is the origin of fibres in R?

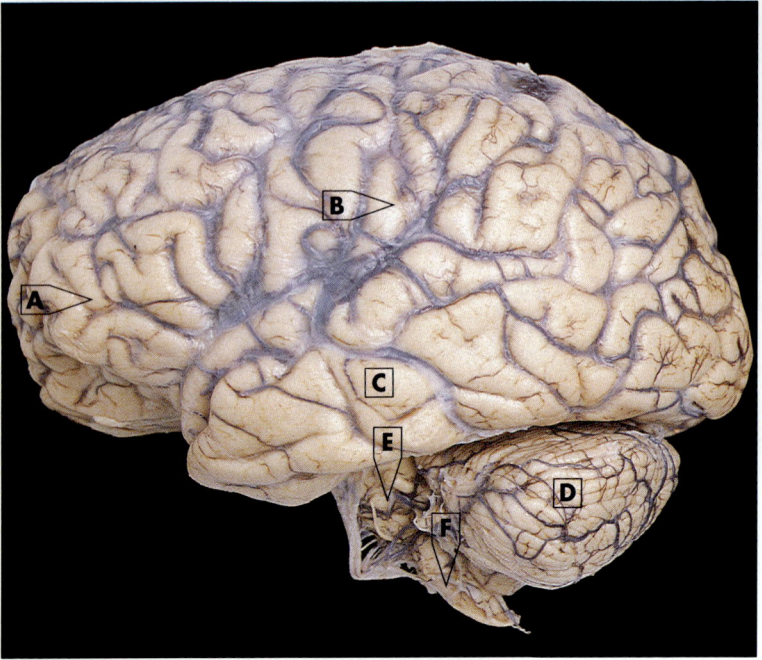

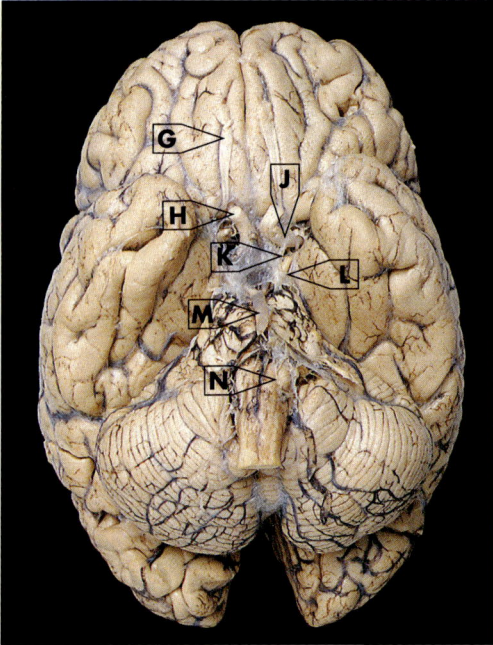

1. In which cranial fossa does A lie?
2. What layer of dura separates C & D?
3. Where do nerves from G leave the cranial cavity?
4. What is the origin of M?

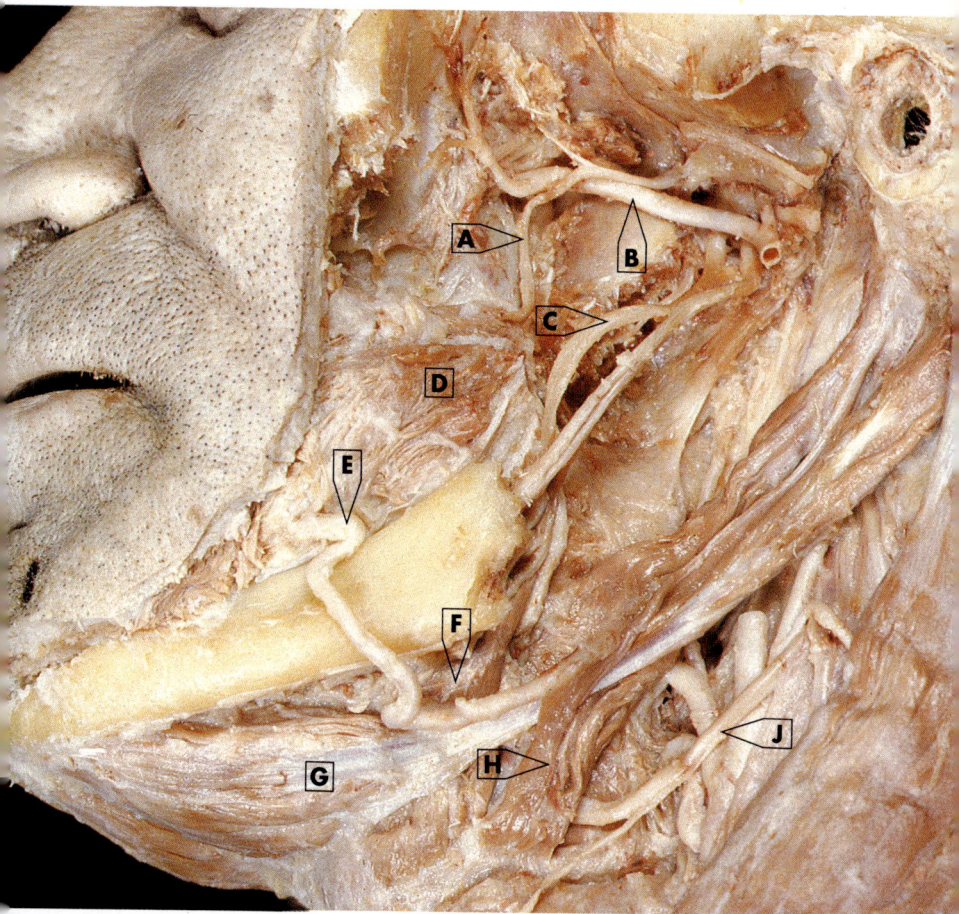

1. What is the distribution of A?
2. What is the action of D?
3. What is the motor supply to F?
4. Where is H attached anteriorly?
5. Which cranial nerve lies deep to H?

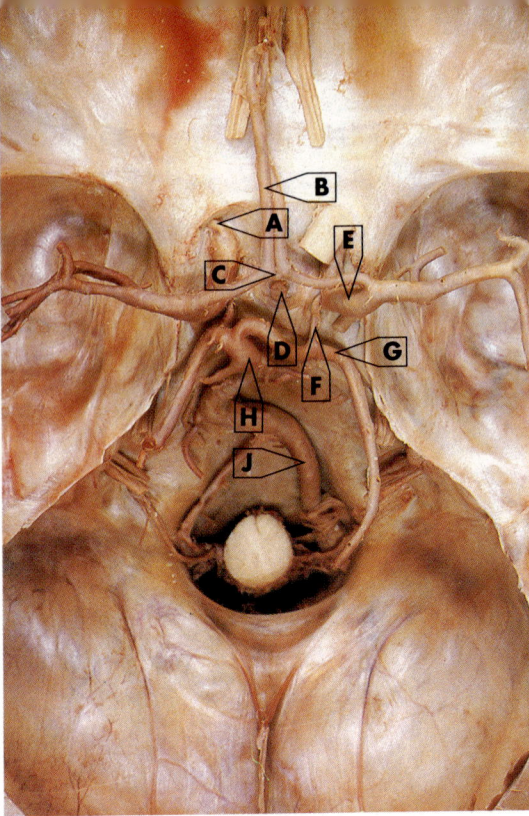

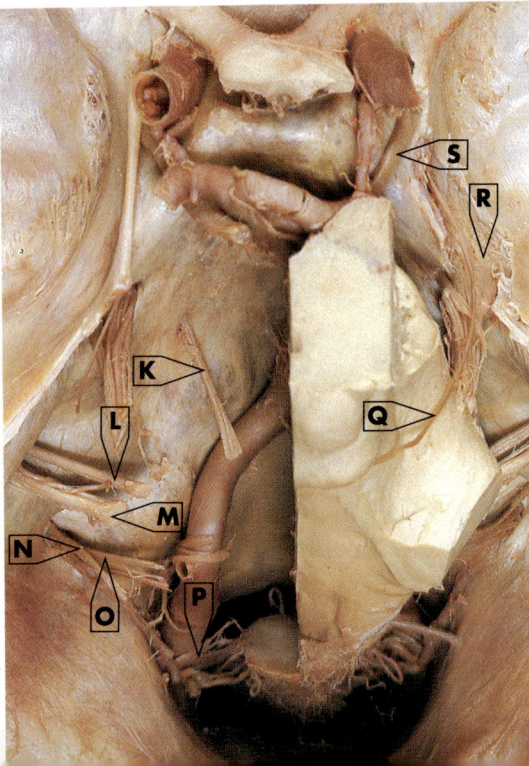

1. What is the distribution of A?
2. What is the distribution of B?
3. How does E enter the cranial cavity?
4. What is the distribution of K?
5. How does L leave the skull?
6. What is the distribution of P?
7. What is the distribution of Q?

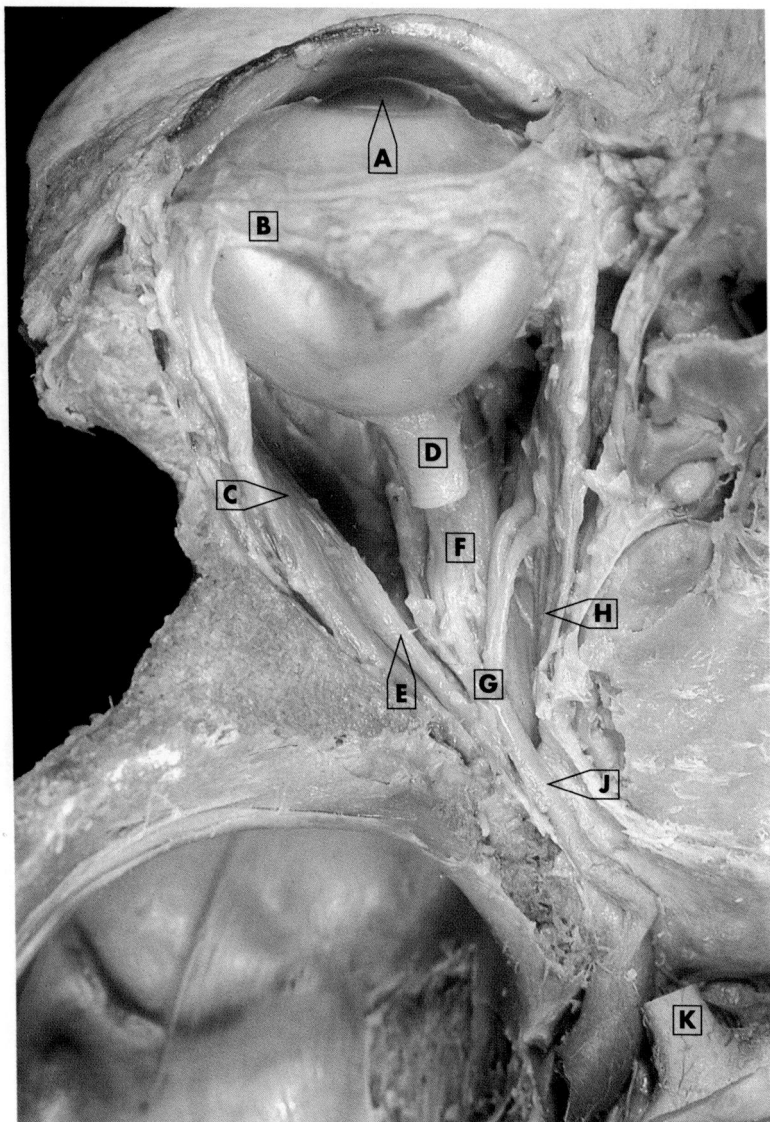

1. What is the action of C?
2. What is the action of H?

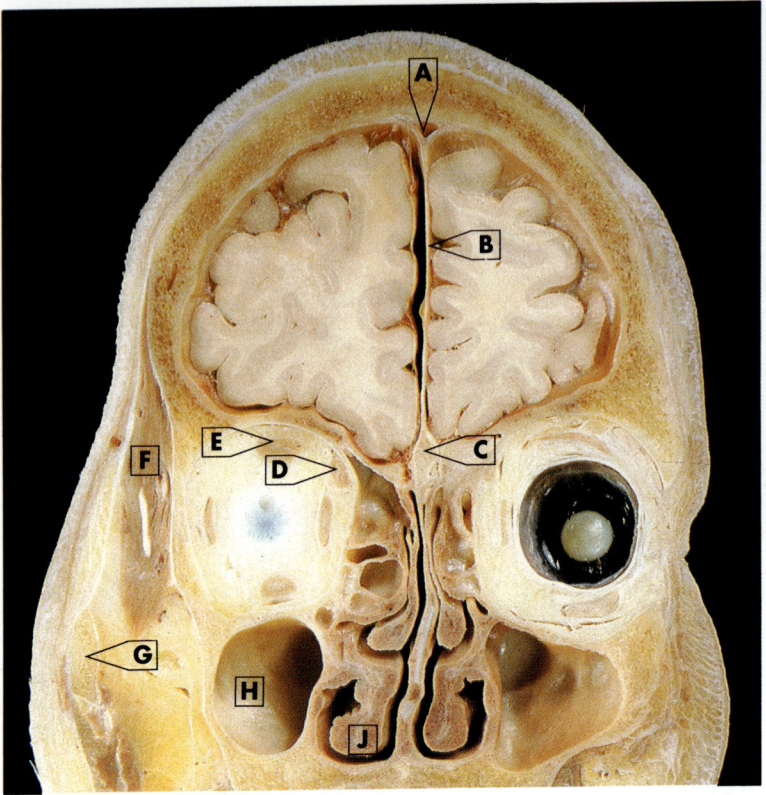

1. Where does A drain?
2. Where does B attach posteroinferiorly?
3. What is the action of D?
4. What is the action of E?
5. Where does H drain?

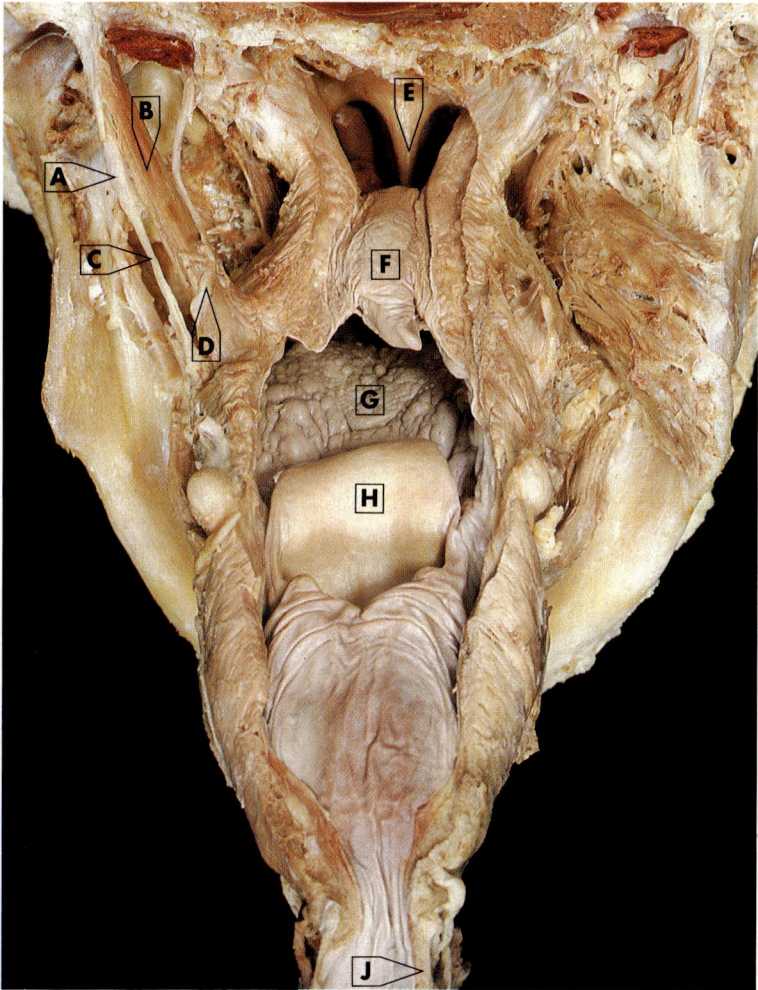

1. What is the motor supply to B?
2. What area lies posterior to E, superior to F?
3. What area lies inferior to F, superior to H and behind G?
4. Which cranial nerve conveys sensation from the area in 3?
5. Which cranial nerve conveys sensation from the area below H?

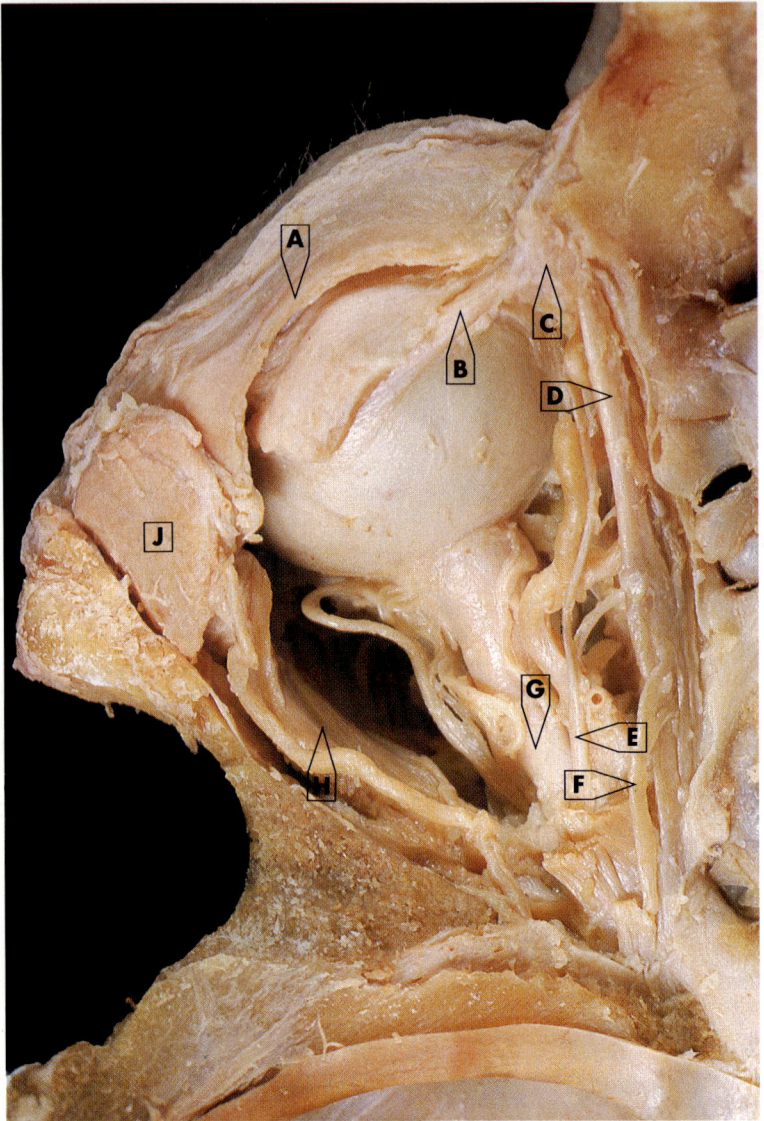

1. What is the motor supply to A?
2. What is the action of D?
3. What is the origin of E?
4. Where does J drain?

5. What is the motor supply to J?

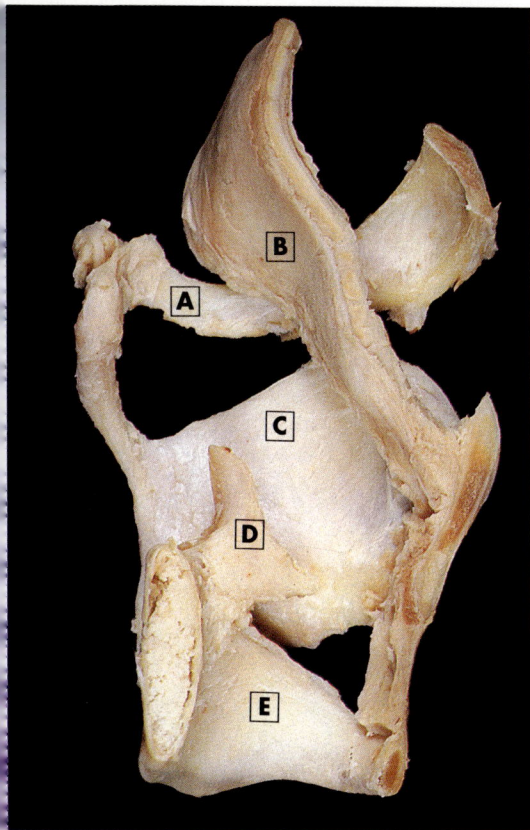

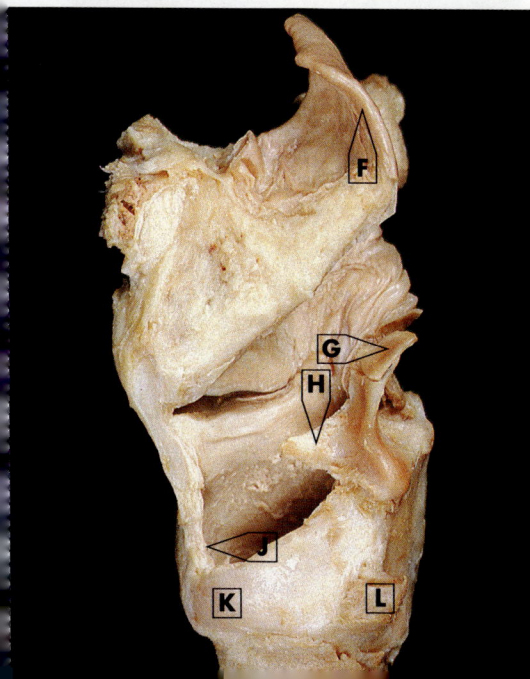

1. What form of cartilage lies in B?
2. Which membrane joins B & D?
3. How does C articulate with E?
4. What is attached at H?

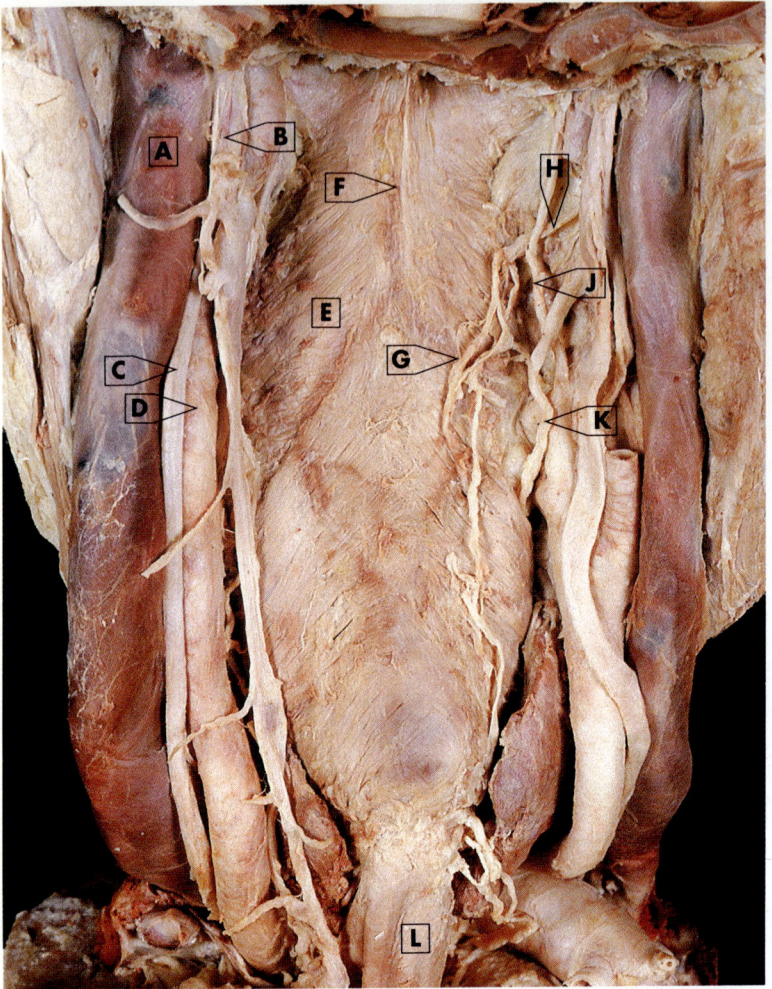

1. What are the anterior attachments of E?
2. Which nerves contribute to G?
3. What is the origin of J?
4. What is the motor distribution of K?
5. What is the motor nerve supply to L?

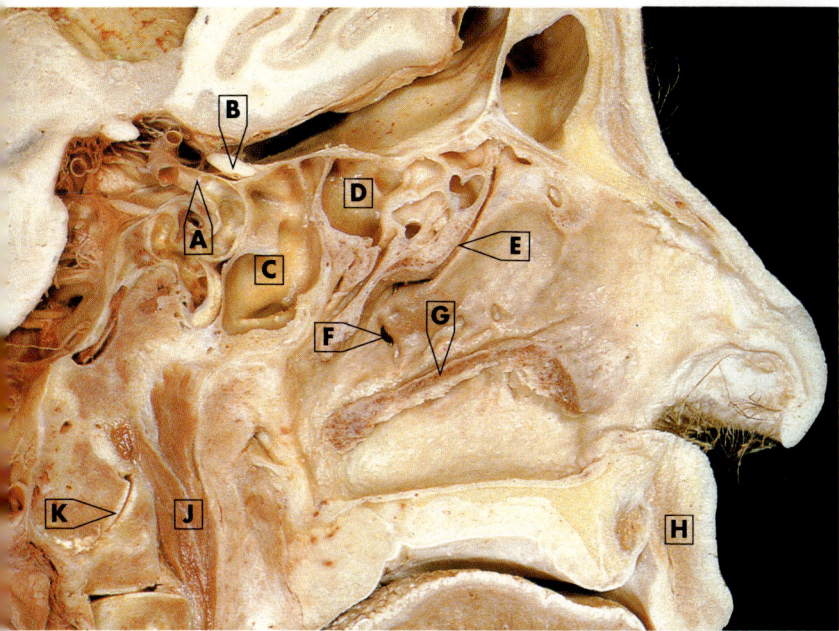

1. Where does C drain?
2. Into which part of the nasal cavity does E open?
3. What opens under G?
4. What is the motor supply to J?
5. What movement takes place at K?

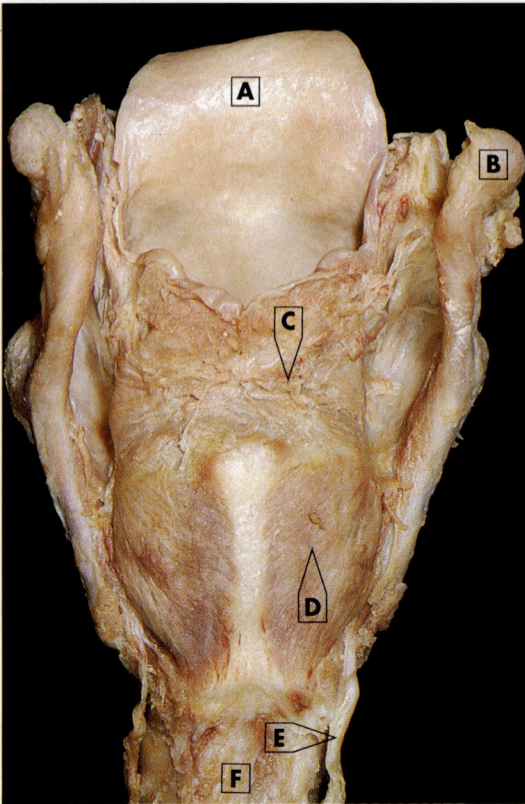

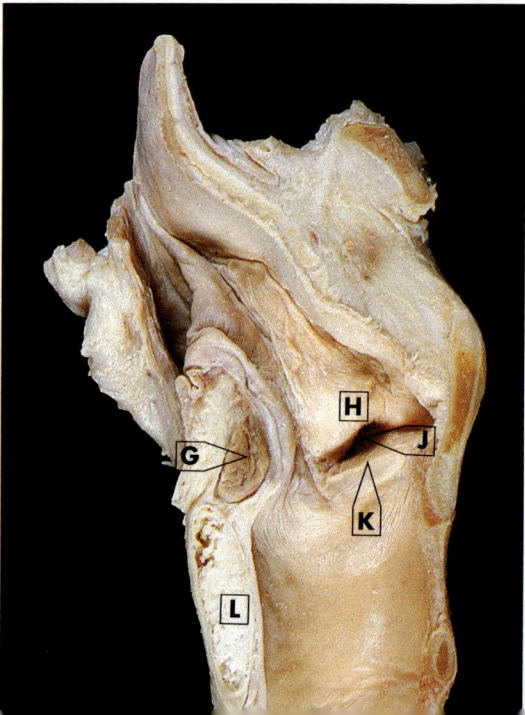

1. What is the action of C?
2. What is the action of D?
3. What is the sensory distribution of E?
4. Which nerve conveys sensation from H?
5. What is the function of J?

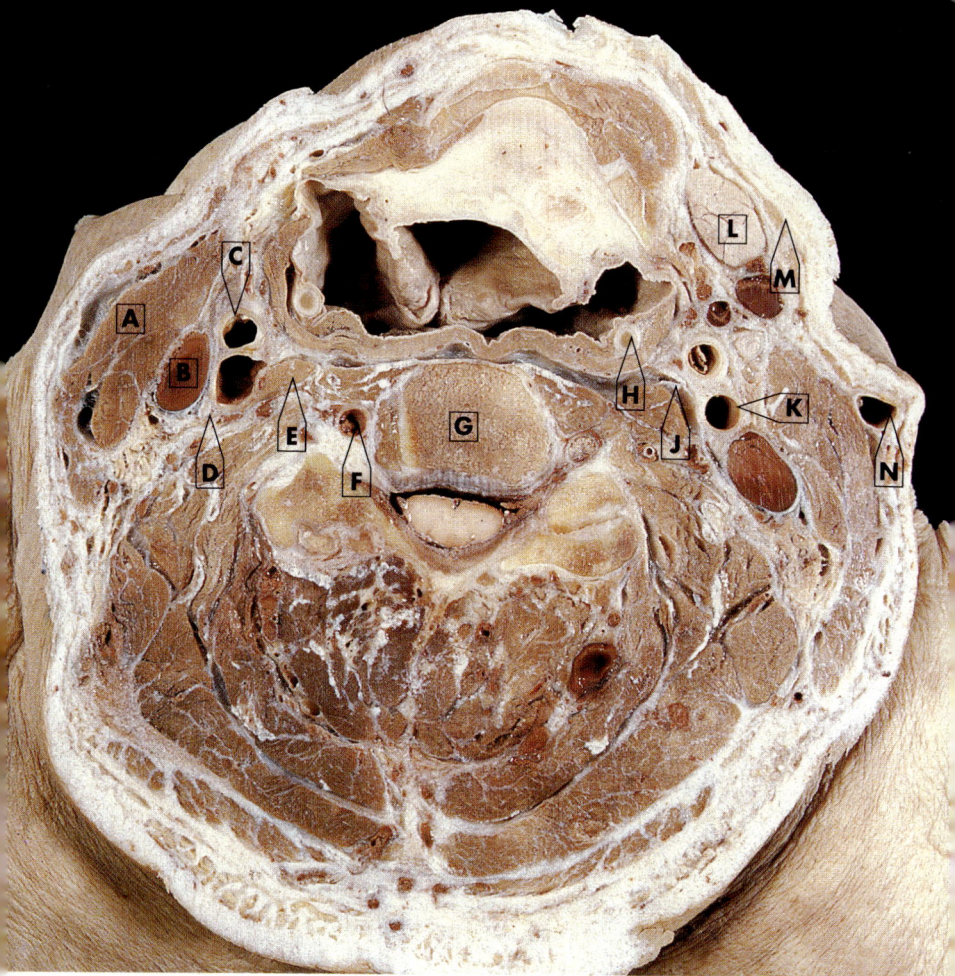

1. Which layer of cervical fascia lies anterior to E?
2. Which layer of cervical fascia encloses B, C & D?
3. What is the origin of F?
4. What is the motor supply to M?
5. Where does N drain?

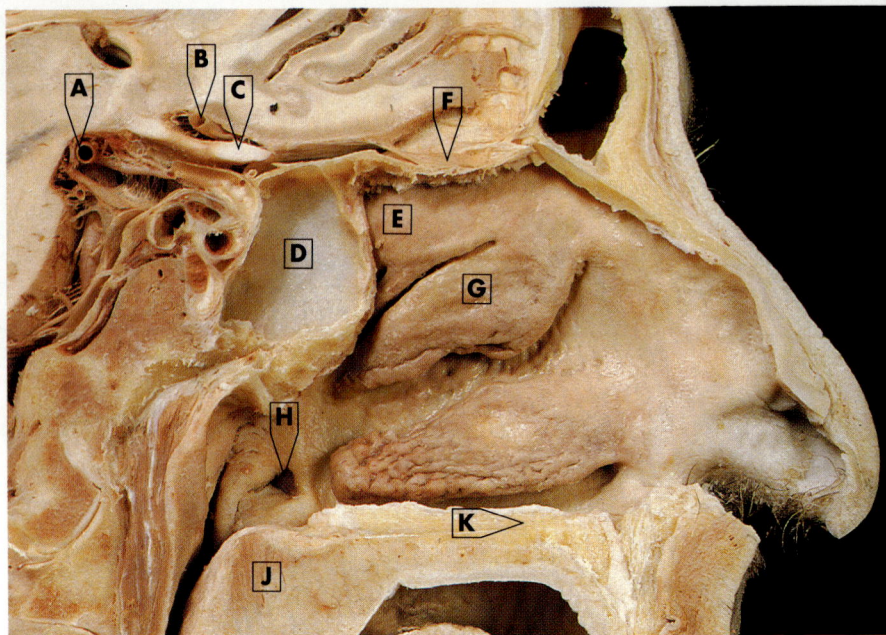

1. What is the origin of A?
2. Which bone contributes to G?
3. What area communicates with the nasopharynx via H?
4. What is the sensory supply to J?
5. Which bone contributes to the anterior part of K?

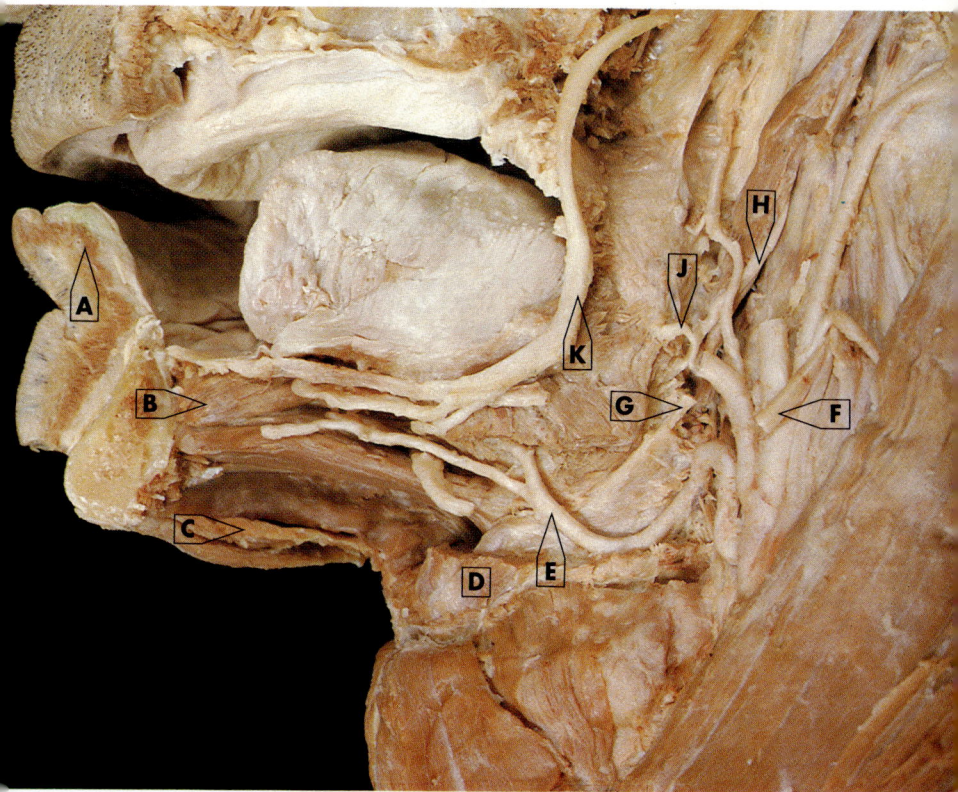

1. What is the action of A?
2. What is the action of B?
3. What is the action of D?
4. What is the origin of E?
5. What is the motor distribution of H?

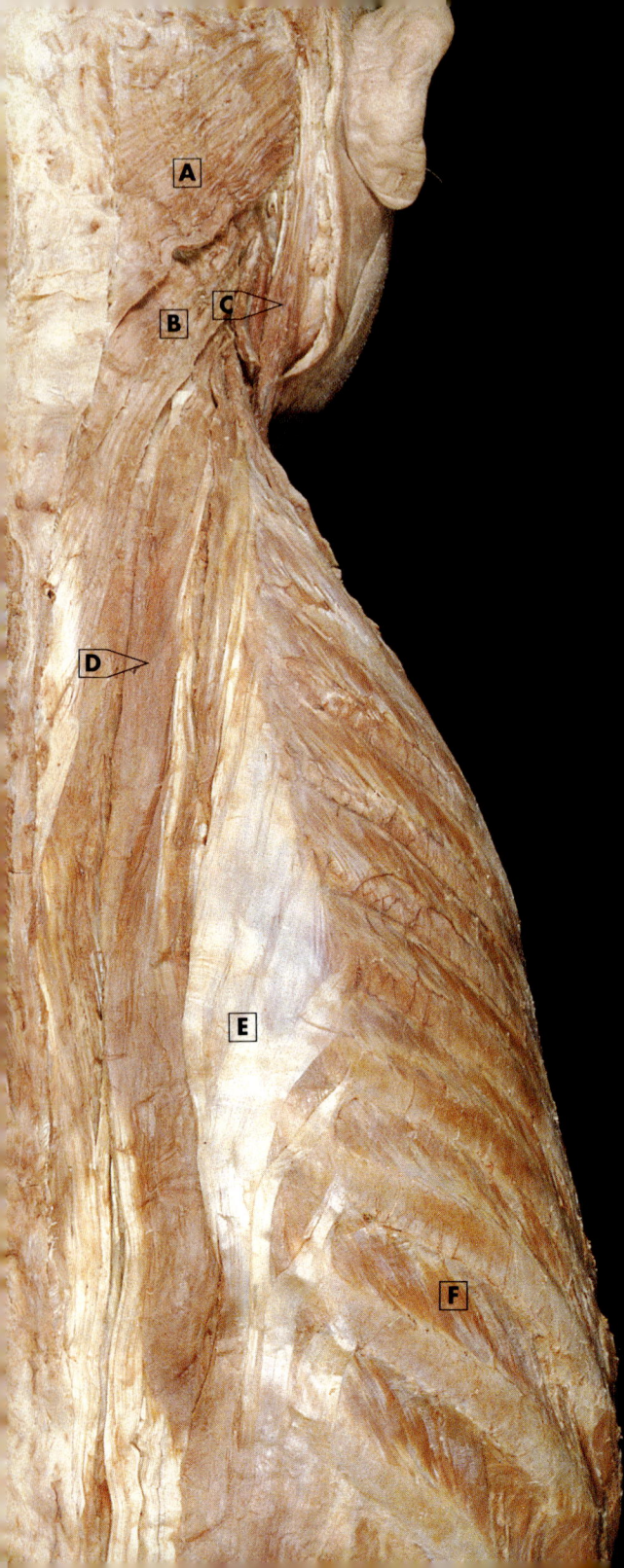

1. Which upper limb muscles have been removed to expose these back muscles?
2. What is the nerve supply of B?
3. Give the function of D.

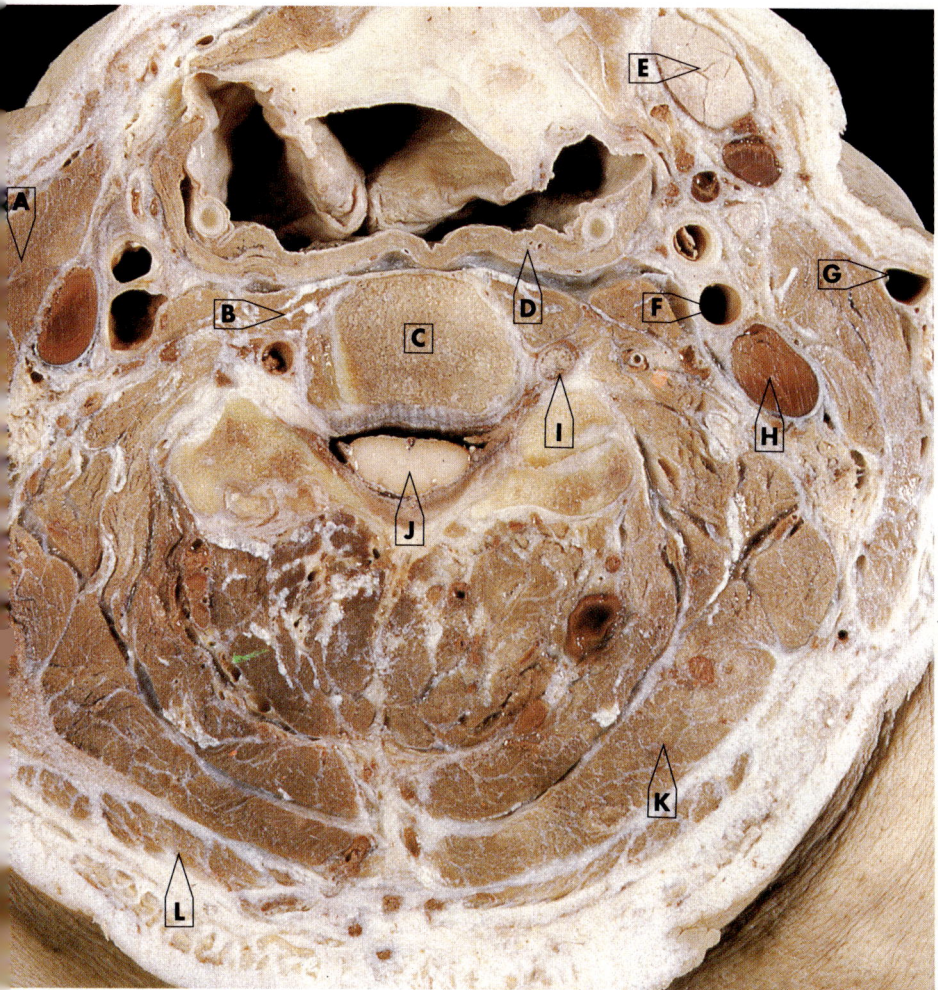

1. What is the level of this section?
2. Which muscles, illustrated here, in addition to deep muscles of the back, are important in controlling head and neck movements?
3. What is the relation of I to the vertebrae?

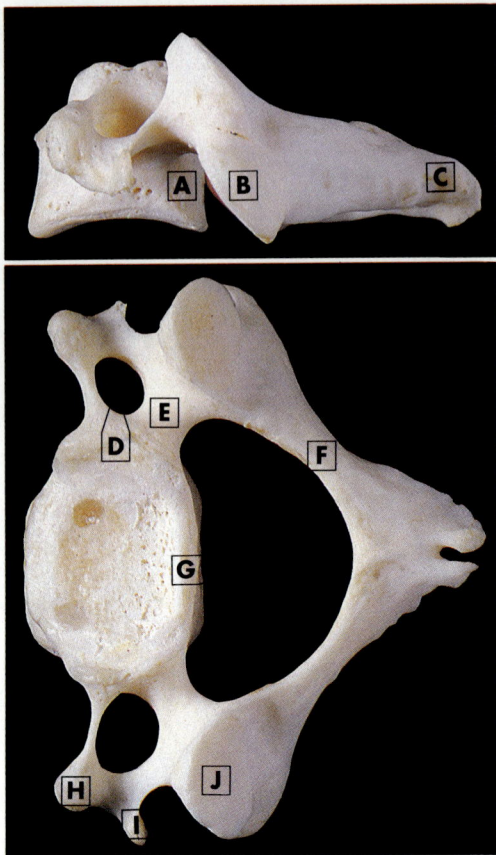

1. What type of joint occurs between B and J?
2. What occupies the groove between H and I?

3. Describe the cervical spinal curve.

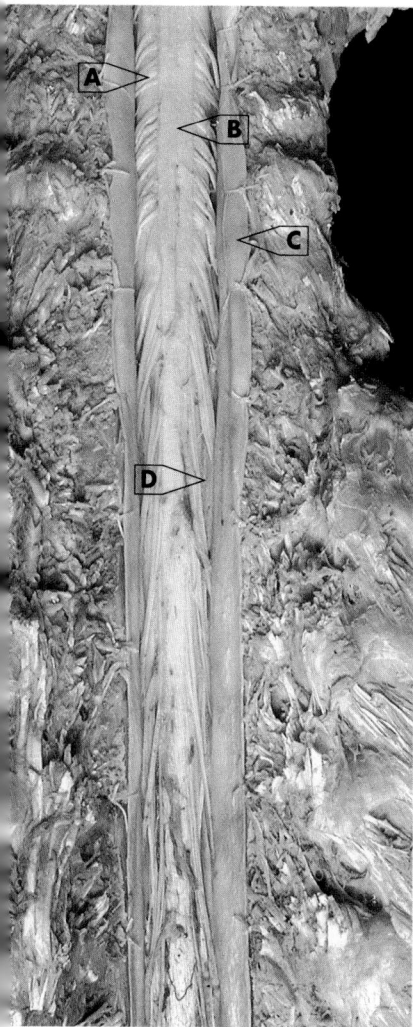

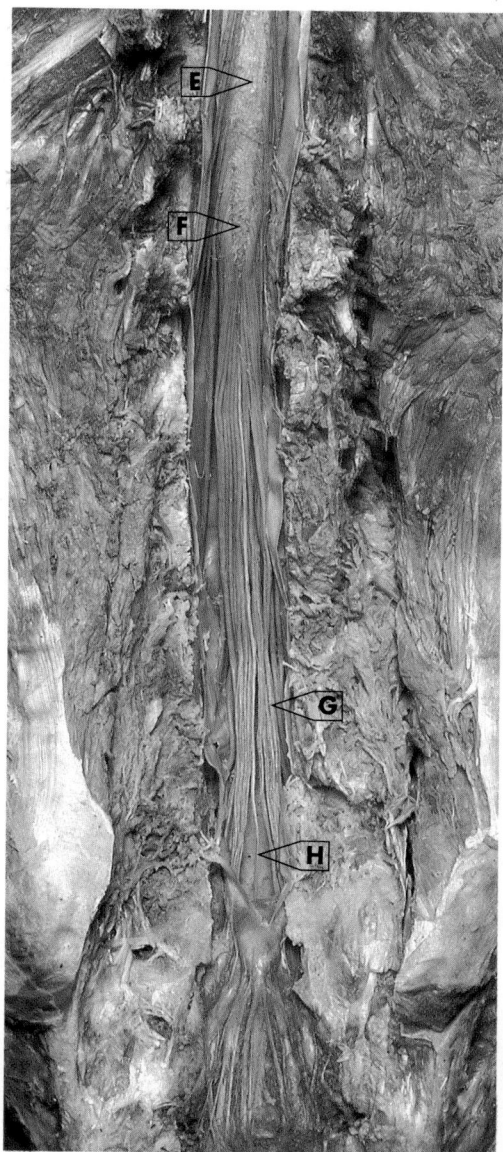

1. Which meningeal layer is closely applied at B?
2. What is the lowest extent of F in an adult?
3. What forms G?

155

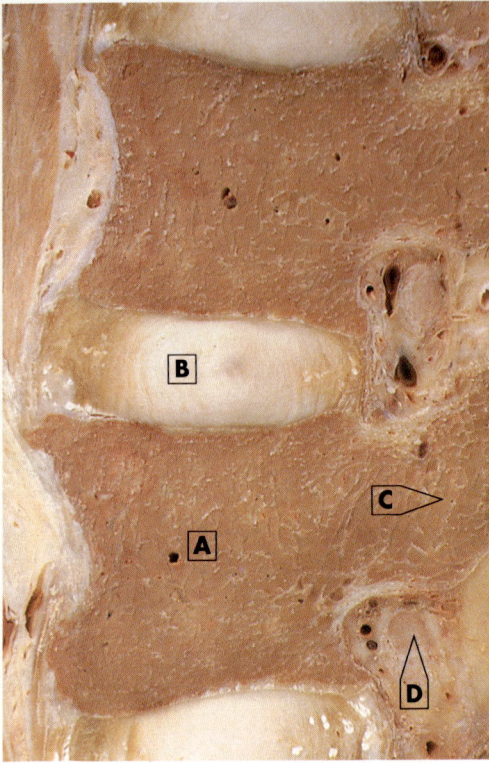

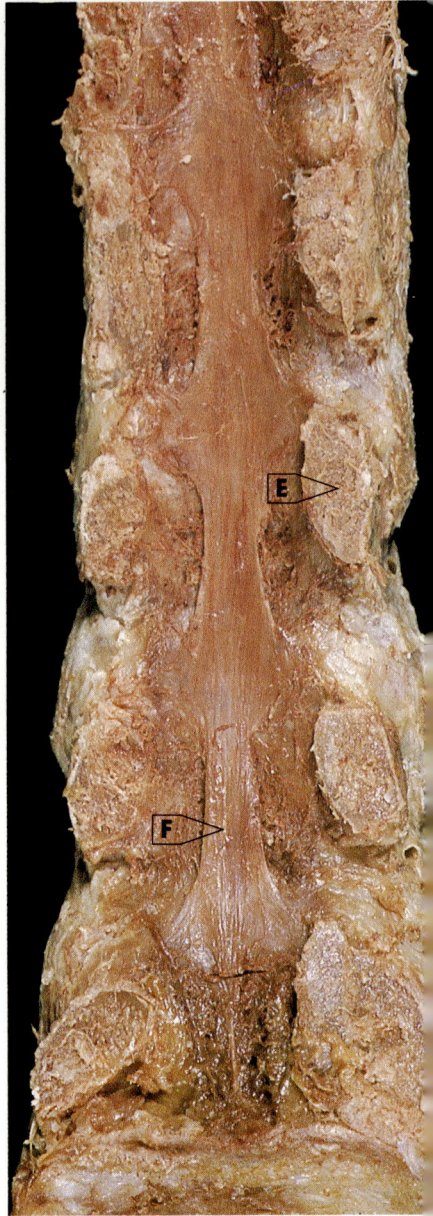

1. What type of joint is B?
2. Give the composition of B.

3. What is the posterior relation of F?

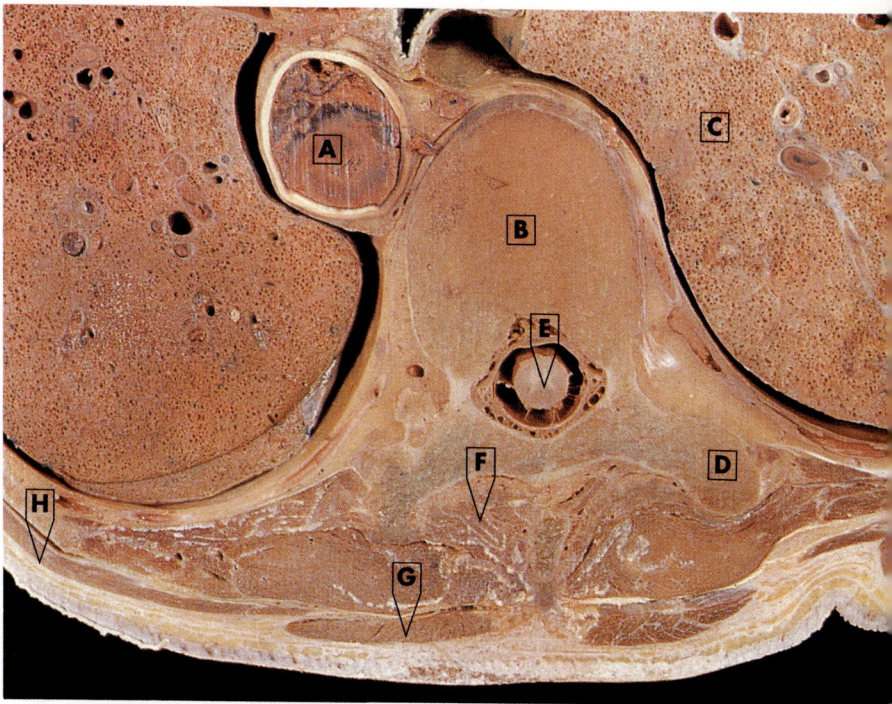

1. What is the level of this section?
2. How is A related to the vertebral column at different levels? 157

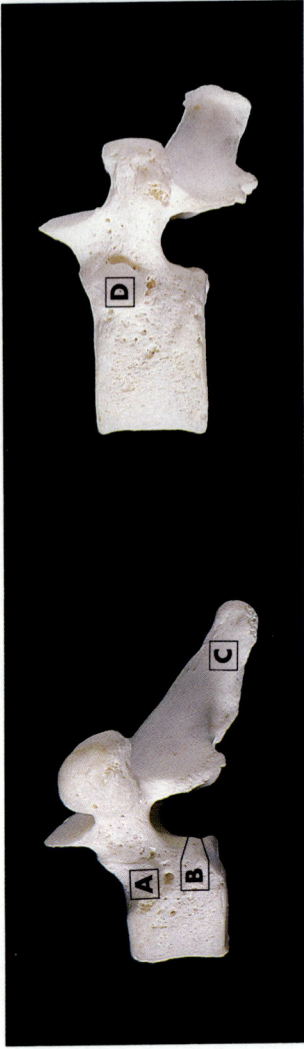

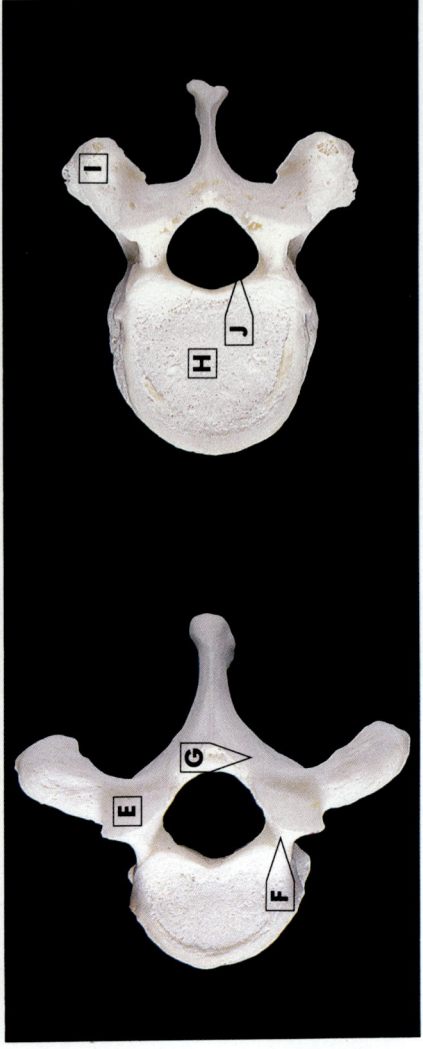

1. Which vertebrae are illustrated?

2. With what does A articulate?

3. What type of bone marrow occupies H?

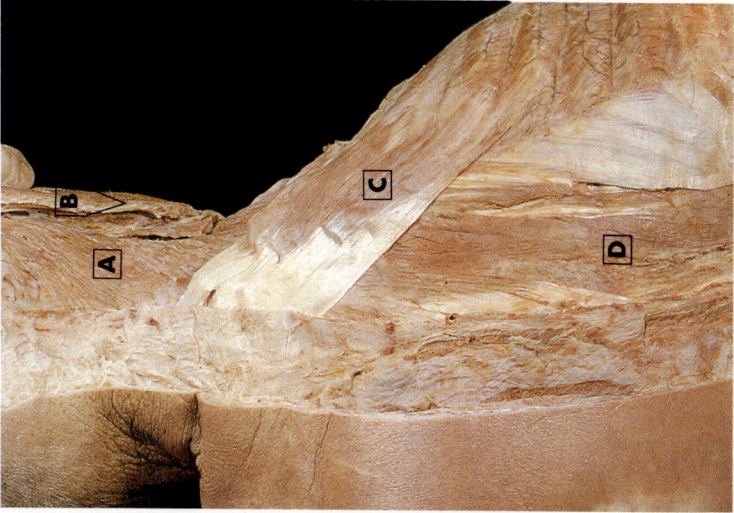

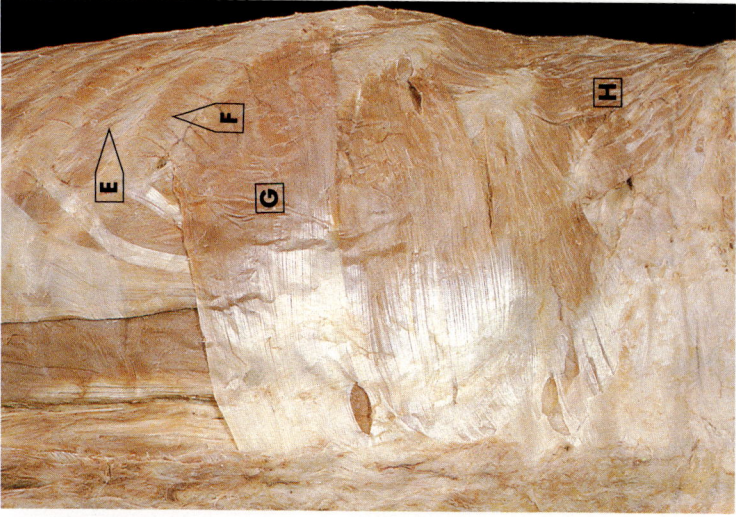

1. What is the function of C and G?

2. Give the nerve supply of C and G.

Captions

Answers

Page 2   **A**   Left vagus X nerve
    B   Left superior intercostal vein
    C   Descending thoracic aorta
    D   Ligamentum arteriosum
    E   Left phrenic nerve
    F   Left brachiocephalic vein
    G   Fibrous pericardium
    H   Left pulmonary artery
    J   Left main bronchus
    K   Left common carotid artery

1. 2nd and 3rd left intercostal spaces.   *2.11*
2. 12th thoracic vertebra.   *2.37*
3. Remnant of ductus arteriosus.   *2.28*
4. Phrenic nerves.   *2.20*
5. Lobar (secondary) bronchi.   *2.18*

Page 3   **A**   Right brachiocephalic vein
    B   Internal thoracic artery
    C   Subclavian artery
    D   Parietal pleura
    E   First rib
    F   Intercostal nerve
    G   Subclavian vein
    H   Parietal pleura
    J   Superior epigastric artery
    K   Left brachiocephalic vein

1. Subclavian artery.   *2.10*
2. Costodiaphragmatic recess.   *2.12*
3. Axillary vein.   *3.10*
4. Intercostal nerves.   *2.11*
5. Superior mediastinum   *2.19*

Page 4   **A**   Left atrium
    B   Left atrial appendage
    C   Superior vena cava
    D   Ascending aorta
    E   Mitral (bicuspid) value
    F   Pulmonary trunk
    G   Right coronary artery
    H   Left atrial appendage

    J   Aortic valve
    K   Right atrial appendage

1. Musculi pectinati.   *2.26*
2. Chordae tendineae.   *2.26*
3. Marginal branch.   *2.29*
4. Prevents backflow into the left ventricle.   *2.22*

Page 5   **A**   Diaphragm
    B   Pericardium (fibrous)
    C   Lingula
    D   Left brachiocephalic vein
    E   Brachiocephalic trunk
    F   Trachea
    G   Horizontal fissure
    H   Cardiac notch
    J   Inferior border of lung
    K   Anterior border of lung

1. Phrenic nerves.   *2.12*
2. Phrenic nerves.   *2.20*
3. Left internal jugular and subclavian veins.   *2.33*
4. Arch of aorta.   *2.33*

Page 6   **A**   Coronary sinus
    B   Pulmonary vein
    C   Left atrium
    D   Superior vena cava
    E   Ascending aorta
    F   Right ventricle
    G   Anterior interventricular artery
    H   Pulmonary trunk
    J   Left ventricle
    K   Circumflex artery

1. Right atrium.   *2.31*
2. Posterior vein of left ventricle, great and middle cardiac veins.   *2.31*
3. Middle (inferior).   *2.19*
4. Middle (inferior).   *2.20*
5. Infundibulum.   *2.24*
6. Anastomosis with right coronary artery.   *2.30*

Page 7 **A** Sternum (body)
B Anterior mediastinum
C Ascending aorta
D Superior vena cava
E Right pulmonary artery
F Pectoralis major
G Right main bronchus
H Oesophagus
J Descending thoracic aorta
K Intervertebral disc

1. Fat, remnants of thymus
gland. *2.19*
2. Arch of aorta. *2.28*
3. Pulmonary trunk. *2.28*
4. Left atrium. *2.26*
5. Abdominal aorta. *4.62*

Page 8 **A** Apex
B Notch for left subclavian artery
C Groove for arch of aorta
D Oblique fissure
E Groove for descending thoracic
aorta
F Pulmonary vein
G Cardiac impression
H Diaphragmatic surface
J Inferior border
K Lingula

1. 2.5cm above medial
third of clavicle. *2.13*
2. Lateral edge of first rib. *3.9*
3. Brachiocephalic trunk, left
common carotid and
left subclavian arteries. *2.33*
4. Pulmonary ligament. *2.13*
5. Left ventricle. *2.21*
6. Costodiaphragmatic
recess. *2.17*

Page 9 **A** Anterior interventricular artery
B Right coronary artery
C Pulmonary trunk
D Pulmonary valve
E Left coronary artery
F Aortic valve
G Superior vena cava
H Left atrial appendage
J Left atrium
K Left superior pulmonary vein

1. Pulmonary sinus. *2.28*
2. Medial end of second left
intercostal space. *2.25*
3. Medial ends of first and
second right intercostal
spaces. *2.27*
4. Circumflex artery. *2.30*
5. Mitral (bicuspid) valve. *2.26*

Page 10 **A** Trachea
B Right vagus (X) nerve
C Right common carotid artery
D Right subclavian artery
E Left vagus (X) nerve
F Azygos vein
G Right main bronchus
H Oesophagus
J Oesophageal plexus
K Descending thoracic aorta

1. Lower part of fourth thoracic
vertebral body. *2.19*
2. Right recurrent laryngeal
nerve. *7.11*
3. Superior vena cava. *2.38*
4. Tenth thoracic vertebra. *2.37*
5. Twelfth thoracic vertebra. *2.37*

Page 11 **A** Left subclavian artery
B Left phrenic nerve
C Ligamentum arteriosum
D Left vagus (X) nerve
E Fibrous pericardium
F Central tendon of diaphragm
G Left atrial appendage
H Left dome of diaphragm

1. Axillary artery. *3.9*
2. Diaphragm. *2.35*
3. Left recurrent laryngeal
nerve. *2.37*
4. Pericardiophrenic artery. *2.20*
5. Phrenic nerve. *4.63*
6. Liver. *4.33*

Page 17 **A** Internal thoracic artery
B Right phrenic nerve
C Right lung root
D Right atrium
E Fibrous pericardium
F Left phrenic nerve
G Pulmonary trunk
H Left lung root
J Right ventricle
K Left ventricle

1. Pulmonary vessels and bronchi. *2.16*
2. Superior and inferior vena cava, coronary sinus, cardiac veins. *2.23*
3. Central tendon of diaphragm. *2.20*
4. Infundibulum of right ventricle *2.24*
5. Tricuspid and pulmonary valves. *2.24, 2.25*

Page 18 **A** Internal jugular vein
B Left common carotid artery
C Left vagus (X) nerve
D Left recurrent laryngeal nerve
E Right vagus (X) nerve
F Azygos vein
G Thoracic duct
H Inferior vena cava
J Descending thoracic aorta
K Oesophagus

1. Superior mediastinum. *2.33*
2. Aortic opening. *2.38*
3. Aortic opening. *2.37*
4. Eighth thoracic vertebra. *4.62*
5. Right phrenic nerve. *2.35*
6. Azygos vein and left gastric vein. *2.37*

Page 19 **A** Ascending aorta
B Right atrium
C Right coronary artery
D Marginal artery
E Great cardiac vein
F Ventricular branch of left coronary artery
G Anterior interventricular artery
H Anterior interventricular artery
J Left phrenic nerve
K Left ventricle

1. Lies between the third and sixth right costal cartilages 3cm from midline. *2.21*
2. Apex of heart. *2.29*
3. Coronary sinus. *2.31*
4. Left shoulder. *4.63*
5. Upwards and medially from apex to the second left intercostal space 3cm from midline. *2.21*

Page 20 **A** Subclavian vein
B Internal jugular vein
C Left brachiocephalic vein
D Left superior intercostal vein
E Left phrenic nerve
F Left vagus (X) nerve
G Right phrenic nerve
H Inferior thyroid vein
J Ligamentum arteriosum
K Right atrial appendage

1. Thoracic duct. *2.37*
2. Second and third posterior intercostal veins. *2.11*
3. Pulmonary trunk (left pulmonary artery). *2.33*
4. Visceral layer of serous pericardium. *2.20*

Page 25 A Hypothenar muscles
B Pisiform
C Ulnar artery
D Flexor retinaculum
E Median nerve
F Tendon of flexor pollicis longus
G Thenar muscles
H Tendon of flexor carpi radialis
I Trapezium
J Intercarpal joint
K Tendon of extensor carpi ulnaris

1. Median nerve.   *3.51*
2. Deep branch of ulnar
   nerve.   *3.24*
3. Articular (hyaline) cartilage. *1.13*

Page 26 A Subclavian artery
B Scalenus medius
C Lower trunk
D Middle trunk
E Upper trunk
F Lateral cord
G Posterior cord
H Medial cord
I Musculocutaneous nerve
J Coracobrachialis
K Subscapularis
L Axillary nerve

M Median nerve
N Axillary artery
O Thoracodorsal nerve
P Subscapular artery
Q Radial nerve
R Ulnar nerve

1. Pectoralis major and
   pectoralis minor.   *3.7*
2. Brachiocephalic trunk.   *2.33*
3. As brachial artery at lower
   border of teres major.   *3.13*

Page 27 A Deep fascia
B Ulnar nerve
C Median nerve
D Brachial artery
E Humerus
F Brachialis
G Biceps brachii
H Cephalic vein
I Radial nerve
J Triceps brachii

1. Musculocutaneous nerve.   *3.13*
2. From medial cord in axilla.   *3.11*
3. 'Wrist drop'.   *3.40*

Page 28 A Radial artery
B Flexor digitorum superficialis
C Tendon of abductor pollicis longus
D Flexor retinaculum
E Ulnar nerve
F Abductor pollicis brevis
G Recurrent branch of median nerve
H Opponens digiti minimi
I Abductor digiti minimi
J Brachialis
K Common flexor origin
L Pronator teres
M Tendon of biceps brachii
N Flexor carpi radialis
O Brachioradialis
P Flexor carpi ulnaris
Q Median nerve

1. Synovial sheaths.   *3.51*
2. Radial nerve (main nerve, not
   posterior interosseous).   *3.37*
3. Hook of hamate via pisiform
   and pisohamate
   ligament.   *3.16*

169

Page 29 A Deltoid
B Scapula
C Infraspinatus
D Subscapularis
E Intercostal muscle(s)
F Rib
G Humerus
H Coracobrachialis and short head of biceps (combined)

I Axillary artery
J Axillary vein
K Axillary fat
L Serratus anterior
M Pectoralis minor
N Pectoralis major
O Superficial fascia

Page 30 A Scalenus anterior
B Supraspinatus
C Posterior cord
D Coracobrachialis
E Tendon of long head of biceps
F Axillary nerve
G Radial nerve
H Subscapularis
I Long thoracic nerve
J Thoracodorsal nerve
K Serratus anterior
L Brachial artery
M Brachialis

1. Posterior divisions of three trunks of brachial plexus. *3.11*
2. Deltoid, teres minor and upper lateral cutaneous nerve of arm. *3.6*
3. Protraction and rotation of scapula. *2.8*

Page 31 A Fibrous flexor sheath
B Third lumbrical muscle
C First lumbrical muscle
D First dorsal interosseous
E Adductor pollicis
F Tendon of flexor pollicis longus
G Abductor digiti minimi
H Opponens digiti minimi
I Tendon of flexor digitorum superficialis
J Tendon of flexor digitorum profundus
K Abductor pollicis brevis
L Tendon of flexor carpi radialis
M Tendon of abductor pollicis longus

1. Deep branch of ulnar nerve. *3.26*
2. Deep branch of ulnar nerve. *3.24*
3. Sides of middle phalanx. *3.16*

Page 32 A Anconeus
B Extensor carpi ulnaris
C Extensor digitorum
D Tendon of extensor digiti minimi
E Flexor carpi ulnaris
F Subcutaneous border of ulna
G Abductor pollicis longus
H Extensor pollicis longus

I Extensor indicis
J Extensor pollicis brevis
K Dorsal tubercle of radius
L Tendon of extensor carpi radialis longus
M Tendon of extensor carpi radialis brevis

Page 33 **A** Clavicle
B Pectoralis major
C Axillary artery
D Axillary fat
E Head of humerus
F Glenoid fossa of scapula
G Serratus anterior
H Blade of scapula
I Subscapularis
J Infraspinatus
K Deltoid

1. Axillary nerve.   3.6
2. Clavicular fibres of pectoralis
   major and anterior fibres
   of deltoid.   3.44
3. Subscapularis bursa.   3.44

Page 34 **A** Superior angle of scapula
B Supraspinatus
C Suprascapular ligament
D Coracoid process
E Capsule of shoulder joint
F Tendon of long head of biceps
G Tendon of latissimus dorsi
H Surgical neck of humerus
I Subscapularis
J Inferior angle

1. Adduction, extension and
   medial rotation of
   shoulder joint.   3.30
2. Axillary nerve.   3.33
3. Lesser tubercle of humerus. 3.31

Page 35 **A** Digital branch of ulnar nerve
B Digital artery
C Digital branch of median nerve
D Palmar artery
E Superficial palmar arch
F Recurrent branch of median nerve
G Ulnar nerve
H Flexor retinaculum
I Ulnar artery
J Median nerve
K Tendon of flexor carpi radialis
L Radial artery

1. Flexor carpi ulnaris and
   part of flexor digitorum
   profundus.   3.17
2. Brachial artery in cubital
   fossa.   3.15
3. As deep palmar arch.   3.18

Page 36 **A** Infraspinatus
B Teres minor
C Axillary nerve
D Inferior angle of scapula
E Teres major
F Long head of triceps
G Radial nerve
H Lateral head of triceps
I Teres major
J Tendon of latissimus dorsi
K Medial head of triceps
L Profunda brachii artery
M Nerve to lateral head of triceps
N Lateral head of triceps (cut and displaced)
O Vena comitans
P Nerve to medial head of triceps
Q Radial nerve

1. Suprascapular nerve.   3.33
2. Medial lip of intertubercular
   sulcus.   3.33
3. Extension of elbow joint.   3.34

Page 41　**A** Abductor pollicis longus
B Extensor carpi radialis brevis
C Extensor pollicis brevis
D Extensor carpi radialis longus
E Radial artery
F Extensor indicis
G Extensor digitorum
H First dorsal interosseous
I Extensor pollicis longus
J First metacarpal

1. Anatomical snuff box.　*3.39*
2. Cephalic vein and branches of superficial radial nerve.　*3.39*
3. Stabilises, extends and abducts wrist.　*3.37*

Page 42　**A** Acromion
B Tendon of supraspinatus
C Coracoacromial ligament
D Tendon of infraspinatus
E Tendon of long head of biceps
F Coracoid process
G Glenoid fossa
H Teres minor
I Glenoid labrum
J Capsule of shoulder joint
K Subscapularis

1. Subacromial bursa.　*3.44*
2. Deepens gleonoid fossa, promoting stability of shoulder joint.　*3.43*
3. Synovial membrane.　*3.44*

Page 43　**A** Digital branch of ulnar nerve
B Digital branch of median nerve
C Abductor digiti minimi
D Palmar aponeurosis
E Adductor pollicis
F Palmaris brevis
G Abductor pollicis brevis
H Pisiform
I Ulnar artery
J Flexor carpi radialis
K Tendon of palmaris longus

1. Aponeurosis of palmaris longus, stabilises skin of palm and protects underlying vessels and nerves.　*3.21*
2. Common flexor origin of humerus.　*3.16*
3. C6, C7 and C8.　*3.4*

Page 44　**A** Brachialis
B Brachioradialis
C Posterior interosseous nerve
D Tendon of biceps
E Pronator teres
F Superficial branch of radial nerve
G Supinator
H Flexor carpi radialis
I Flexor digitorum superficialis

1. Musculocutaneous nerve.　*3.13*
2. Flexes elbow joint and rotates forearm to position between pronation/supination.　*3.37*
3. Brachioradialis and extensor carpi radialis longus.　*3.37*

Page 49 | A | Sternohyoid
| B | Clavicle
| C | Subclavius
| D | Sternothyroid
| E | Interclavicular ligament
| F | Articular disc
| G | First rib
| H | Manubrium
| I | First costal cartilage
| J | Parietal pleura

1. In membrane initially, not from cartilage model.
2. Stabilises sternoclavicular joint. *3.42*
3. Pectoralis major, intercostal muscles. *2.10*

Page 50 | A | Basilic vein
| B | Flexor carpi ulnaris
| C | Ulnar nerve
| D | Flexor digitorum profundus
| E | Flexor digitorum superficialis
| F | Median nerve
| G | Flexor pollicis longus
| H | Brachioradialis
| I | Extensor carpi radialis longus and brevis
| J | Deep fascia
| K | Subcutaneous border of ulna
| L | Shaft of ulna
| M | Interosseous membrane
| N | Radius

1. Median and ulnar nerves. *3.19*
2. Through head of radius and styloid process of ulna. *3.48*
3. Abductor pollicis brevis, flexor pollicis brevis, opponens pollicis; first and second lumbricals. *3.26*

Page 51 | A | Deltoid
| B | Cephalic vein
| C | Clavicle
| D | External jugular vein
| E | Clavicular fibres of sternomastoid
| F | Manubrial fibres of sternomastoid
| G | Tendon of pectoralis major
| H | Branch of thoracoacromial artery
| I | Axillary vein
| J | Pectoralis minor

1. Deltoid tuberosity on lateral surface of shaft of humerus. *3.13*
2. Lateral margin of dorsal venous arch. *3.39*
3. Protracts and rotates scapula, accessory muscle of inspiration. *2.7*

Page 52 | A | Round ligament of liver
| B | Falciform ligament
| C | Fibrous pericardium
| D | Central tendon of diaphragm
| E | Left lobe of liver
| F | Anterior surface of stomach
| G | Greater omentum attaching to greater curvature of stomach
| H | Posterior surface of stomach
| J | Transverse mesocolon
| K | Transverse colon

1. Left umbilical vein. *4.32*
2. Subphrenic space. *4.21*
3. Lesser sac (omental bursa). *4.21*

Page 53
A   Sternocostal head of pectoralis major
B   Serratus anterior
C   Costochondral junction of 8th rib
D   External intercostal muscle
E   Internal intercostal muscle
F   Transversus abdominis
G   Subcostal nerve

1. Adducts and flexes arm at shoulder joint, medially rotates humerus, depresses scapula, accessory muscle of inspiration.   2.7
2. Medial and lateral pectoral nerves from brachial plexus.   2.7
3. Protracts scapula and assists trapezius in scapular rotation during abduction of upper limb.   2.8
4. Long thoracic nerve from upper three roots (C3, 4, 5) of brachial plexus.   2.8
5. T12.   4.59

Page 54
A   Right kidney
B   Liver
C   Bile duct
D   Portal vein
E   Hepatic artery
F   Inferior vena cava
G   (common) Hepatic artery
H   Disc between vertebrae T12 and L1
J   Left suprarenal gland
K   Spleen
L   Gastrosplenic ligament
M   Greater curvature of stomach

1. At greater duodenal papilla in second part of duodenum.   4.28, 4.34
2. Formed by union of splenic and superior mesenteric veins.   4.45
3. Opening into lesser sac (epiploic foramen).   4.21
4. Secondary cartilaginous joint.   1.12, 8.10

Page 55
A   Inferior vena cava
B   Left suprarenal gland
C   Right renal artery
D   Left renal vein
E   Right gonadal artery
F   Inferior vena cava
G   Inferior mesenteric artery
H   Genitofemoral nerve
J   Ureter
K   Left common iliac vein

1. Lesser sac and stomach.   4.23
2. Left kidney, suprarenal gland and gonad.   4.49
3. To cremaster muscle and to superficial tissues of femoral triangle.   4.60, 6.6
4. To gut from left colic flexure to anal canal.   4.43, 5.6

Page 56
A   Descending thoracic aorta
B   Inferior vena cava
C   Anterior vagal trunk
D   Diaphragm
E   Oesophagus
F   Lesser omentum
G   Body of stomach (anterior surface)
H   Greater omentum

1. Posterior intercostal arteries, and arteries to bronchi and oesophagus.   2.37
2. Mostly from left vagus nerve.   4.56
3. Left phrenic nerve.   4.63, 2.35
4. Lesser sac (omental bursa). 4.21

Page 57
A Transverse colon
B Middle colic artery
C Superior mesenteric vein
D Superior mesenteric artery
E Jejunal artery
F Ileocolic artery
G Marginal artery
H Vermiform appendix
J Caecum

1. Posterior to neck of pancreas it joins splenic vein to form portal vein. *4.45*
2. Iliacus and psoas. (Usually the appendix is retrocaecal.) *4.41*
3. Caecum, ileum, urinary bladder, right ureter, right ovary. *4.41*

Page 58
A Right crus of diaphragm
B Sympathetic trunk
C First lumbar nerve
D Tendon of psoas minor
E Disc between vertebrae L4 and 5
F Lateral cutaneous nerve of thigh
G Genitofemoral nerve
H Femoral nerve
J Obturator nerve
K Psoas fascia covering psoas major
L Iliac fascia covering iliacus

1. Descends into pelvic cavity and fuses with its fellow on anterior surface of coccyx. *5.20*
2. L2 and 3. *4.60*
3. Iliacus, sartorius, pectineus, rectus femoris and three vasti. *4.60, 6.10*
4. Sensory branches to hip and knee joints and to skin on medial side of thigh; motor branches to obturator externus, pectineus, gracilis, and adductors longus, brevis and magnus. *6.6, 6.14*

Page 59
A Left lobe
B Left triangular ligament
C Fissure for ligamentum venosum
D Caudate lobe
E Hepatic vein
F Inferior vena cava
G Bare area
H Inferior layer of coronary ligament

1. Stomach. *4.33*
2. Diaphragm. *4.34*
3. Superior recess of lesser sac. *4.34*
4. Lesser omentum. *4.34*

Page 60
A Parietal peritoneum of anterior abdominal wall
B Ridge formed by medial umbilical ligament
C Mucosal lining of urinary bladder
D Psoas major
E Iliacus
F Iliac crest
G Head of femur
H Anterior surface of right kidney
J Polar artery
K Renal vein
L Branch of renal artery
M Ureter

1. Segmental innervation by lower thoracic spinal nerves. *4.17*
2. Medial umbilical ligament, lying deep to this peritoneal ridge, is remnant of umbilical artery. *4.9*
3. Anterior rami of upper lumbar nerves. *4.59*
4. Flexes hip joint. *4.59*
5. Anterior to right kidney lie descending duodenum, right colic flexure and coils of jejunum. *4.48*

Page 61

A   Parietal pleura
B   A right posterior intercostal artery
C   Thoracic duct
D   Azygos vein
E   Descending thoracic aorta
F   Right crus of diaphragm
G   Cysterna chyli
H   Coeliac trunk
J   Superior mesenteric artery
K   A right lumbar artery
L   Sympathetic trunk

1. In root of neck at confluence of left internal jugular and subclavian veins. 2.37, 7.10
2. In superior vena cava.   2.38
3. Hepatic, splenic and left gastric arteries.   4.25
4. Derivatives of mid-gut: part of pancreas, most of small intestine and large intestine as far as left colic flexure.   4.4

Page 62

A   Superior epigastric vessels
B   Transversus abdominis
C   Internal oblique
D   External oblique
E   Arcuate line
F   Parietal peritoneum
G   Inferior epigastric vessels
H   Urachus
J   Medial umbilical ligament
K   Urinary bladder

1. Rectus abdominis.   4.8, 4.10
2. From external iliac artery.   4.10
3. It is the remnant of the allantois.   4.9
4. C and D.   4.9

Page 63

A   Iliolumbar ligament
B   Right gonadal vein
C   Right gonadal artery
D   Superior mesenteric artery
E   Left suprarenal vein
F   Left gonadal vein
G   Ureter
H   Inferior mesenteric artery
J   Bifurcation of aorta

1. Visceral.   4.50
2. Superior rectal artery. 4.43, 5.6
3. Fourth lumbar.   4.50

Page 64

A   Costodiaphragmatic recess
B   Descending thoracic aorta
C   Oesophagus
D   Base of pericardium
E   Inferior vena cava
F   Right dome of diaphragm

1. Right and left vagus nerves. 2.37
2. Left atrium.   2.26
3. Subphrenic space and right lobe of liver.
     4.63, 4.32, 4.21

Page 65

A   Gall bladder
B   Transverse colon
C   Descending duodenum
D   Head of pancreas
E   Superior mesenteric vein
F   Inferior mesenteric vein
G   Body of pancreas
H   Commencement of jejunum

1. Cystic artery, from hepatic artery.   4.35
2. Uncinate process of pancreas.   4.30
3. Distal part of transverse colon, all of descending and sigmoid colon and rectum, upper part of anal canal.   4.44

Page 66
A Testicular artery
B Vas (ductus) deferens
C Pampiniform plexus
D Parietal layer of tunica vaginalis
E Head of epididymis
F Appendix of testis
G Testis
H Cavity of tunica vaginalis
J Hepatic vein
K Inferior vena cava
L Right suprarenal gland
M Superior suprarenal artery
N Right crus of diaphragm
O Middle suprarenal artery
P Renal artery
Q Ureter
R Psoas major
S Inferior vena cava
T Superior mesenteric artery

1. From abdominal aorta just below renal artery. *4.51*
2. In ejaculatory duct. *5.17*
3. Hydrocele. *4.14*
4. Liver and inferior vena cava. *4.34, 4.49*
5. Inferior vena cava and renal veins. *4.48*

Page 67
A Anterior border of spleen
B Fundus of stomach
C Abdominal oesophagus
D Oesophageal branch of left gastric artery
E Left gastric vein
F Caudate lobe of liver
G Hepatic artery
H Portal vein
J Gastroduodenal artery
K Pancreas (neck)
L Antrum of stomach
M Body of stomach

1. In portal vein. *4.45*
2. Right gastroepiploic and superior pancreaticoduodenal arteries. *4.35*
3. Lesser sac and lesser omentum. *4.24*
4. Anterior abdominal wall. *4.23*
5. Opening into lesser sac and inferior vena cava. *4.21 (for Fig 4.36)*

Page 68
A Subcostal nerve
B Quadratus lumborum
C First lumbar nerve
D Lateral cutaneous nerve of thigh
E Iliacus
F Femoral nerve
G Lumbosacral trunk
H Psoas major
J Obturator nerve
K Lumbosacral intervertebral disc
L A lumbar artery

1. It stabilizes 12th rib during inspiration. *4.59*
2. L4 and 5. *4.60*
3. Flexion and medial rotation of hip joint, anterior and lateral flexion of lumbar spine. *5.59, 6.40*

Page 69　**A**　Superior mesenteric artery
　　　　　B　Head of pancreas
　　　　　C　Second part of duodenum
　　　　　D　Liver
　　　　　E　Perirenal fat
　　　　　F　Quadratus lumborum
　　　　　G　Right renal vein
　　　　　H　Abdominal aorta
　　　　　J　Psoas major
　　　　　K　Ureter
　　　　　L　Left kidney
　　　　　M　Spleen

1. From front of abdominal aorta
　　　at level of vertebra L1. *4.38*
2. Renal fascia. *4.48*
3. In urinary bladder, at ureteric
　　　orifice. *5.12*
4. No. *4.26*

Page 70　**A**　Ascending colon
　　　　　B　Ileocaecal valve
　　　　　C　Appendicular artery
　　　　　D　Mucosa of caecum
　　　　　E　Mesoappendix
　　　　　F　Appendix
　　　　　G　Sigmoid colon
　　　　　H　Fibrous pericardium
　　　　　J　Right lobe of liver
　　　　　K　Left lobe of liver
　　　　　L　Greater omentum
　　　　　M　Transverse colon
　　　　　N　Jejunum

1. From iliocolic artery. *4.43*
2. It joins rectum in front of
　　　third sacral vertebra. *4.43*
3. Right ventricle. *2.22*
4. It is displaced upwards
　　　towards liver and
　　　stomach, to which it is
　　　adherent. *4.18*

Page 71　**A**　Portal vein
　　　　　B　Superior mesenteric artery
　　　　　C　Splenic vein
　　　　　D　Splenic artery
　　　　　E　Inferior mesenteric vein
　　　　　F　3rd part of duodenum
　　　　　G　Superior mesenteric vein
　　　　　H　First part of duodenum, displaced

1. Near porta hepatis it divides
　　　into left and right
　　　branches. *4.45*
2. To pancreas, spleen and
　　　stomach. *4.27*
3. More commonly it joins
　　　splenic vein. *4.44*
4. Inferior vena cava and
　　　abdominal aorta. *4.28*

Page 72　**A**　Oesophagus
　　　　　B　Inferior vena cava
　　　　　C　Left lobe of liver
　　　　　D　Lesser omentum
　　　　　E　Body of stomach
　　　　　F　Spleen
　　　　　G　Costodiaphragmatic recess
　　　　　H　Right gastroepiploic artery
　　　　　J　Fundus of gall bladder
　　　　　K　External oblique aponeurosis
　　　　　L　Ilioinguinal nerve
　　　　　M　Fascia lata
　　　　　N　Suspensory ligament of penis
　　　　　O　External spermatic fascia
　　　　　P　Great saphenous vein

1. Vagus nerves. *2.37, 4.25*
2. Left and right gastric
　　　arteries and veins. *4.25*
3. To skin of medial side of upper
　　　thigh, and to parts of
　　　external genitalia. *4.10, 6.6*
4. Venous arch on dorsum
　　　of foot. *6.4, 6.32*

Page 73 A Mucosa of superior duodenum
B Liver
C Head of pancreas
D Major duodenal papilla
E Plica circularis in descending
   duodenum
F Lacunar ligament
G Inferior epigastric artery
H Vas (ductus) deferens
J Testicular vessels
K Femoral nerve

L External iliac artery
M External iliac vein
N Obturator nerve
O Genitofemoral nerve
P Ureter

1. Duodenal bulb or cap. *4.28*
2. Bile and main pancreatic
   ducts. *4.31, 4.34*
3. L2, L3 and L4. *4.60*
4. Femoral ring. *6.10, 4.13*

Page 74 A Fundus of gall bladder
B Left branch of hepatic artery
C Portal vein
D Bile duct
E Gastroduodenal artery
F Third part of duodenum
G Uncinate process of pancreas
H Superior mesenteric artery
J Splenic vein
K Tail of pancreas
L Anterior border of spleen
M Left kidney

1. Where lateral edge of
   rectus abdominis (linea
   semilunaris) crosses
   costal margin. *4.35*
2. To left, quadrate and
   most of caudate lobes
   of liver. *4.35*
3. First part of duodenum. *4.29*
4. It lies within splenorenal
   (lienorenal) ligament. *4.31*

Page 75 A Central tendon of diaphragm
B Inferior vena cava
C Inferior phrenic artery
D Oesophagus
E Coeliac trunk
F Right crus of diaphragm
G Superior mesenteric artery
H Renal artery
J Vertebrocostal trigone

1. Fibrous pericardium and
   diaphragmatic surface of
   heart. *4.63*
2. Formed by confluence of left
   and right common iliac veins
   at level of fifth lumbar
   vertebra. *4.52, 4.53*
3. Vagal trunks. *4.62*
4. Lower intercostal nerves. *4.63*

Page 76 A Inferior mesenteric artery
B Sympathetic trunk
C Ureter
D Exteral iliac artery
E Internal iliac artery
F Vas (ductus) deferens
G Median sacral artery
H Common iliac artery
J Common iliac vein
K Testicular vessels
L Genitofemoral nerve
M Femoral nerve

1. It is the artery to the hindgut:
   descending and sigmoid
   colon, rectum and anal
   canal. *4.4*
2. It passes behind inguinal
   ligament to enter
   thigh as femoral
   artery. *4.52, 6.10*
3. Inferior epigastric
   artery. *4.52, 4.10*
4. To most pelvic organs (except
   ovary and rectum), to plevic
   walls and floor, perineum,
   buttock and thigh. *5.22*

Page 77
A Greater omentum (displaced)
B Transverse colon
C Jejunum
D Mesentery of small intestine
E Parietal peritoneum
F Caecum
G Sigmoid colon
H Urinary bladder
J Spermatic cord

1. Transverse colon and greater curvature of stomach. *4.19*
2. Jejunal and ileal arteries and veins and accompanying lymph vessels. *4.36*
3. Lower thoracic and first lumbar spinal nerves. *4.2, 4.17*
4. Colon possesses taenia coli and appendices epiploicae. *4.40*

Page 78
A Abdominal aorta
B Inferior mesenteric vein
C Inferior pole of kidney
D Spleen
E Left colic flexure
F Descending branch of left colic artery
G Descending colon
H Common iliac artery

1. Inferior phrenic, lumbar and median sacral arteries. *4.51*
2. Colon and jejunum. *4.48*
3. Yes. *4.43*
4. It bifurcates in front of sacroiliac joint to form external and internal iliac arteries. *4.52*

Page 79
A Renal vein
B Ureter
C Renal pelvis
D Major calix
E Renal papilla
F Renal cortex
G Right hepatic duct
H Left hepatic duct
J Cystic artery
K Cystic duct
L Common hepatic duct
M Bile duct
N Portal vein
O Coeliac trunk

1. Duodenum, coils of small intestine, possibly caecum and appndix. *4.49*
2. Descends behind first part of duodenum and enters second part. *4.34*
3. Traverses free border of lesser omentum. *4.45*
4. It is the artery to the foregut. *4.4*

Page 80
A Renal fascia
B Suprarenal gland
C Renal vein
D Psoas fascia covering psoas major
E Perirenal fat
F Ureter
G Gonadal vein
H Aortic lymph node
J Splenic artery
K Superior mesenteric artery
L Inferior mesenteric artery
M Aortic lymph node
N Bifurcation of aorta

1. Left renal vein, uncinate process of pancreas and third part of duodenum. *4.38*
2. Third part of duodenum. *4.43*
3. To cisterna chyli and thoracic duct. *4.55*

182

Page 81
A Pectoralis major
B Serratus anterior
C External oblique
D Rectus abdominis
E Tendinous intersection
F Transversus abdominis
G Inferior epigastric artery (and venae comitantes)
H Pyramidalis
J Spermatic cord

1. Powerfully flexes lumbar spine, also assists other abdominal muscles in raising intra-abdominal pressure.   *4.9*
2. Yes.   *4.8*
3. From external iliac artery.   *4.10*
4. Vas deferens and nerves and vessels of testis and epididymis.   *4.14*

Page 82
A Anterior superior iliac spine
B Obturator nerve
C Obturator membrane
D Inferior ramus of pubis
E Sacrotuberous ligament
F Sacrospinous ligament
G Piriformis
H Superior gluteal artery
J Anterior ramus of 1st sacral nerve

1. Greater sciatic foramen.   *5.19*
2. Medial compartment of thigh.   *5.21, 6.5, 6.14*
3. Obturator internus.   *5.19*
4. From internal iliac artery.   *5.23*

Page 83
A Corpus spongiosum
B Ischiocavernosus
C Crus of penis
D Perineal membrane
E Levator ani
F Sacrotuberous ligament
G Coccyx
H Gluteus maximus

1. Expands to form glans penis.   *5.27*
2. Perineal nerve, a branch of pudendal nerve.   *5.26*
3. Deep perineal pouch containing external urethral sphincter and bulbourethral glands.   *5.27*
4. Ischiorectal fossa.   *5.26*

Page 84
A Bladder (detrusor muscle)
B Prostatic urethra
C Ampulla of rectum
D Membranous urethra
E Anal canal
F Intrabulbar fossa
G Spongy urethra
H Glans penis
J Navicular fossa

1. Parasympathetic supply to detrusor muscle of bladder from pelvic splanchnic nerves via pelvic plexus of autonomic nerves.   *5.13*
2. Membranous urethra is surrounded by external urethral sphincter, lateral to which are medial borders of levator ani muscles.   *5.14*
3. Internal anal sphincter (smooth) and external anal sphincter (striated).   *5.24*
4. Pudendal nerve.   *5.15*

Page 93
A	Femoral nerve
B	Psoas major
C	Vas deferens
D	Ureter
E	Left common iliac vein
F	Rectovesical pouch
G	Sigmoid colon
H	Ureteric orifice
J	Internal iliac artery
K	Ureter
L	External iliac vein
M	Superior vesical artery
N	Obturator artery
O	Obturator nerve
P	Uterine artery
Q	Urachus

1. Anterior rami lumbar nerves 2, 3 and 4.   4.60
2. Anterior to 3rd piece of sacrum it is continuous with rectum.   5.6
3. Ureter passes close to lateral aspect of cervix of uterus.   5.11
4. Urachus is remnant of allantois.   5.12

Page 94
A	Tibialis anterior
B	Extensor digitorum longus
C	Peroneus brevis
D	Tibialis posterior
E	Soleus
F	Medial head of gastrocnemius
G	Anterior tibial artery
H	Deep peroneal nerve
J	Posterior tibial artery
K	Tibial nerve

1. Dorsiflexion of ankle joint, extension of all toes except hallux.   6.34
2. Peroneus longus, common peroneal nerve and its terminal branches.   6.36
3. Flexor compartment.   6.24
4. During locomotion (as opposed to standing) when a wide range of movement is required.   6.24
5. Supplies skin on adjacent sides of great and second toes.

Page 95
A	Superficial circumflex iliac vein
B	Superficial epigastric vein
C	Great saphenous vein
D	Superficial external pudendal vein
E	Femoral vein
F&G	Superficial inguinal lymph nodes

1. Umbilicus.   6.7
2. Saphenous opening.   6.7
3. External iliac vein.   6.10
4. Perineum, buttock, anterior abdominal wall below umbilicus, entire lower limb except lateral side of leg and foot.   6.7
5. Into deep inguinal nodes in femoral canal.   6.7

Page 96
A	Gluteus medius
B	Gluteus maximus
C	Piriformis
D	Tensor fasciae latae
E	Superior gluteal artery
F	Inferior gluteal artery
G	Sciatic nerve
H	Inferior gluteal nerve
J	Posterior cutaneous nerve of thigh
K	Fascia lata

1. Prevents pelvic tilt to unsupported side during walking, abducts and weakly extends hip.   6.16
2. Laterally rotates hip.   6.18
3. From internal iliac artery.   5.23
4. L4, L5; S1, S2, S3.   6.18

186

Page 97
  A Superficial peroneal nerve
  B Great saphenous vein
  C Deep peroneal nerve
  D Dorsal vein arch
  E Common peroneal nerve
  F Anterior tibial artery
  G Superficial peroneal nerve
  H Peroneus longus
  J Extensor digitorum longus
  K Tibialis anterior
  L Extensor hallucis longus

1. In femoral vein in groin.   *6.7*
2. Foot-drop and anaesthesia
    of skin of anterior lower leg
    most of dorsum of foot. *6.36*
3. On dorsum of foot as dorsalis
    pedis artery.   *6.36*

Page 98
  A Sartorius
  B Iliopsoas
  C Pectineus
  D Adductor longus
  E Gracilis
  F Tensor fasciae latae
  G Vastus lateralis
  H Rectus femoris
  J Vastus medialis
  K Quadriceps tendon

1. Flexion and medial rotation
    of hip, lateral and
    anterior flexion of
    lumbar spine.   *4.59*
2. Adduction and medial
    rotation of hip. *6.14*
3. Quadriceps femoris. *6.8*
4. L2, L3, L4.   *6.10*
5. To base of patella. *6.8*

Page 99
  A Peroneus longus tendon
  B Attachment of tibialis posterior to navicular
  C Tibialis anterior tendon
  D Long plantar ligament – completing tunnel for peroneus longus tendon
  E Long plantar ligament
  F Peroneus brevis tendon
  G One of the tarsal attachments of tibialis posterior
  H Medial calcaneal tubercle
  J Lateral calcaneal tubercle
  K Flexor hallucis longus tendon

1. Eversion of foot, weak
    plantar flexion of ankle
    joint.   *6.36*
2. Inversion of foot, weak
    plantar flexion of
    ankle joint. *6.25*
3. Stabilize calcaneo-cuboid
    joint, support lateral
    longitudinal arch. *6.56*
4. Plantar aponeurosis, flexor
    digitorum brevis,
    abductor hallucis,
    flexor accessorius.
       6.26, 6.27, 6.28
5. Flexor surface, base of distal
    phalanx of hallux. *6.25*

Page 100
  A Femoral nerve
  B Femoral artery
  C Femoral vein
  D Femoral canal
  E Psoas bursa
  F Superior pubic ramus
  G Obturator externus
  H Adductor magnus
  J Ilio-psoas tendon
  K Capsule of hip joint

1. It lies outside the femoral
    sheath.   *6.10*
2. Femoral artery is within the
    sheath.   *6.10*
3. Femoral ring.   *6.10*
4. Obturator canal. *5.21*
5. To the lesser trochanter. *4.59*

Page 104 A   Ilio-psoas
    B   Pectineus
    C   Obturator externus
    D   Adductor brevis
    E   Gracilis
    F   Adductor magnus
    G   Obturator nerve
    H   Anterior division of obturator nerve
    J   Posterior division of obturator nerve
    K   Sartorius
    L   Popliteal artery
    M   Popliteal vein
    N   Quadriceps tendon
    O   Suprapatellar bursa
    P   Capsule of knee joint

Q   Hamstring muscles
R   Infrapatellar fat pad
S   Popliteus
T   Meniscus
U   Gastrocnemius

1. Lumbar nerves L1, L2 and L3.   *4.60*
2. Pectineus, gracilis, adductors longus and brevis.   *6.14*
3. Intracapsular, extrasynovial.   *6.45*
4. Lateral rotation of femur on tibia.   *6.46*

Page 105 A   Distal phalanx of great toe
    B   Head of third metatarsal
    C   Middle phalanx of second toe
    D   Styloid process of fifth metatarsal
    E   Tuberosity of navicular
    F   Facets for tendon of peroneus longus
    G   Facets for tendon of tibialis anterior
    H   Groove inferior to cuboid
    J   Groove inferior to sustentaculum tali
    K   Head of talus

1. Plantar plate of fibrocartilage.   *6.54*
2. Tendon of tibialis posterior. *6.25*
3. Tendon of peroneus longus   *6.31*
4. Tendon of flexor hallucis longus.   *6.27*

Page 106 A   Flexor hallucis brevis
    B   Flexor accessorius
    C   Tendon of flexor hallucis longus
    D   Plantar metatarsal artery
    E   Second lumbrical
    F   Plantar arch
    G   Stump of flexor digitorum brevis
    H   Abductor hallucis
    J   Tendon of flexor digitorum longus
    K   Abductor digiti minimi

1. Flexes proximal phalanx of hallux.   *6.28*
2. Flexes lateral four digits irrespective of position of ankle joint.   *6.28*
3. Styloid process at base of fifth metatarsal.   *6.31*
4. Medial plantar nerve.   *6.27*

Page 107 A   Obturator internus
    B   Adductor magnus
    C   Gluteus maximus
    D   Long head of biceps femoris
    E   Short head of biceps femoris
    F   Semitendinosus
    G   Semimembranosus
    H   Common peroneal nerve
    J   Tibial nerve
    K   Popliteal artery
    L   Rectus femoris
    M   Vastus medialis
    N   Sartorius
    O   Adductor longus
    P   Adductor magnus

Q   Semitendinosus
R   Femoral artery
S   Great saphenous vein
T   Components of sciatic nerve

1. Linea aspera and lateral supracondular ridge of femur.   *6.20*
2. Sensory to knee joint, lateral cutaneous nerve of calf, lateral sural cutaneous nerve.   *6.22*
3. Prevents lateral displacement of patella.   *6.8*
4. Extends hip, flexes knee.   *6.19*

Page 108 A Femoral vein
    B Femoral artery
    C Femoral nerve
    D Pectineus
    E Iliopsoas
    F Sartorius
    G Obturator internus
    H Acetabular labrum
    J Sciatic nerve
    K Gluteus maximus

1. Saphenous nerve.   *6.6*
2. Anterior division of obturator nerve, femoral nerve (frequently).   *6.14*
3. Flexes and medially rotates hip joint: forward and lateral flexion of lumbar spine.   *4.59*
4. Gluteal tuberosity of femur, iliotibial tract.   *6.15*

Page 109 A Tibial tubercle
    B Tibialis anterior
    C Superficial peroneal nerve
    D Extensor retinaculum
    E Extensor digitorum longus
    F Tendon of extensor hallucis longus
    G Muscle belly of extensor digitorum brevis
    H Tendon of extensor hallucis brevis
    J Tendon of peroneus tertius
    K Dorsalis pedis artery
    L Talocalcaneal joint
    M Medial collateral ligament
    N Ankle joint
    O Tendon of tibialis posterior

P Tendon of flexor digitorum longus
Q Tendon of flexor hallucis longus
R Abductor hallucis
S Flexor digitorum brevis
T Abductor digiti minimi
U Tendon of peroneus brevis

1. Dorsiflexion and inversion of foot.   *6.34*
2. Deep peroneal nerve.   *6.36*
3. Synovial saddle joint.   *6.52*
4. Stabilises ankle, prevents forward displacement of tibia on talus.   *6.51*

Page 110 A Femoral nerve
    B Iliacus
    C Gluteal muscles
    D Articular cartilage of acetabulum
    E External iliac artery
    F Ligamentum teres
    G Obturator internus
    H Obturator externus
    J Adductor muscles
    K Vastus lateralis

1. L2, L3 and L4.   *6.10*
2. By iliopsoas tendon to lesser trochanter.   *4.59*
3. Limits adduction of hip joint.   *6.40*
4. Through lesser sciatic foramen to greater trochanter.   *6.18*

Page 111 A Tendon of flexor hallucis longus
    B Tendon of flexor digitorum longus
    C Tendon of flexor digitorum brevis
    D Medial head of flexor hallucis brevis
    E Flexor digitorum longus
    F Flexor accessorius
    G Abductor hallucis
    H Lateral plantar artery
    J Lateral plantar nerve
    K Medial plantar nerve

1. Base of distal phalanx.   *6.25*
2. Flexion of middle and proximal phalanges.   *6.27*
3. Posterior tibial artery, deep to flexor retinaculum.   *6.31*
4. Abductor hallucis, flexor digitorum brevis, first lumbrical, flexor hallucis brevis.   *6.31*

Page 112 A   Flexor retinaculum
B   Tendon of tibialis posterior
C   Tendon of flexor digitorum longus
D   Tendon of flexor hallucis longus
E   Tendon of flexor digitorum longus
F   Abductor hallucis
G   Medial plantar artery
H   Medial plantar nerve
J   Lateral plantar nerve
K   Lateral plantar artery

1. Medial maleolus and medial surface of calcaneum.   6.23
2. Flexion of toes and plantar flexion of ankle joint.   6.25
3. Medial plantar nerve.   6.27
4. Plantar arch.   6.31

Page 113 A   Gluteus minimus
B   Gluteus medius
C   Tensor fasciae latae
D   Deep branch of superior gluteal artery
E   Superior gluteal nerve
F   Piriformis
G   Greater trochanter
H   Sciatic nerve
J   Inferior gluteal nerve
K   Posterior cutaneous nerve of thigh

1. Abducts and medially rotates hip joint.   6.16
2. Superior gluteal nerve.   6.16
3. Between anterior superior iliac spine and iliac tubercle.   6.16
4. Sacral nerves S1, S2 and S3.   6.18

Page 114 A   Capsule of hip joint
B   Anterior division of obturator nerve
C   Posterior division of obturator nerve
D   Obturator externus
E   Hamstrings
F   Adductor magnus
G   Gracilis
H   Femoral vein
J   Femoral canal
K   Great saphenous vein
L   Nerve to vastus medialis
M   Saphenous nerve

1. Hip joint and skin on medial side of thigh.   6.14
2. Adducts hip, flexes and medially rotates knee.   6.14
3. A femoral hernia may descend through it.   6.10
4. To the medial side of the first metatarsophalangeal joint.   6.32

Page 115 A   Lateral malleolus
B   Tendo calcaneus
C   Anterior talofibular ligament
D   Calcaneofibular ligament
E   Posterior talofibular ligament
F   Posterior articular surface of calcaneum
G   Capsule of ankle joint
H   Trochlear surface of talus

1. Short saphenous vein and sural nerve.   6.51
2. Gastocnemius and soleus.   6.24
3. Forced inversion may stretch or tear it (sprained ankle).   6.51
4. Inferior aspect of talus.   6.93

Page 116 A   Plantar arch
  B   Perforating branch to dorsum of foot
  C   Deep transverse metatarsal ligament
  D   Fibrous flexor sheath
  E   Dorsal interosseous muscle
  F   Plantar interosseous muscle
  G   Plantar metatarsal artery
  H   Adductor hallucis
  J   Flexor hallucis longus tendon
  K   Medial plantar artery

1. From lateral plantar artery. *6.31*
2. Interosseous tendons. *6.31*
3. Extends interphalangeal and flexes metatarsophalangeal joint, abducts digit. *6.31*
4. Lateral plantar nerve. *6.28*

Page 117 A   Lateral femoral condyle
  B   Posterior cruciate ligament
  C   Anterior cruciate ligament
  D   Medial collateral ligament
  E   Lateral collateral ligament
  F   Tendon of popliteus
  G   Lateral meniscus
  H   Medial meniscus
  J   Transverse ligament

1. It prevents lateral displacement of patella during knee extension. *6.46*
2. Stabilizes knee, prevents posterior displacement of femur on tibia. *6.46*
3. Lateral femoral epicondyle. *6.45*
4. By their anterior and posterior horns to the intercondylar eminence. *6.45*

Page 118 A   Ankle joint
  B   Subtaloid (posterior talocalcaneal) joint
  C   Interosseous talocalcaneal ligament
  D   Medial calcaneal tubercle
  E   Talonavicular joint
  F   Abductor hallucis
  G   Plantar calcaneonavicular (spring) ligament
  H   Plantar aponeurosis
  J   Flexor hallucis longus tendon
  K   Sesamoid bone

1. Dorsiflexion and plantarflexion. *6.50*
2. Inversion and eversion. *6.53*
3. Medial plantar nerve. *6.27*
4. Supports head of talus. *6.53*

Page 119 A   Piriformis
  B   Obturator internus and gemelli
  C   Quadratus femoris
  D   Sacrotuberous ligament
  E   Pudendal nerve
  F   Nerve to obturator internus
  G   Nerve to quadratus femoris
  H   Perineal branch of posterior cutaneous nerve of thigh
  J   Inferior gluteal artery
  K   Internal pudendal artery

1. Through greater sciatic foramen. *6.18*
2. Laterally rotates hip. *6.18*
3. Inferior rectal, perineal and dorsal nerve of penis (or clitoris). *5.26*
4. From internal iliac artery. *5.23*

Page 120 A   Great (long) saphenous vein
       B   Tibialis anterior tendon
       C   Anterior tibial artery
       D   Extensor hallucis longus tendon
       E   Peroneus brevis tendon
       F   Sural nerve
       G   Flexor hallucis longus tendon
       H   Tibial nerve
       J   Posterior tibial artery
       K   Tibialis posterior tendon

1. Medial side of dorsal venous arch.   *6.57*
2. Lateral border of little toe.   *6.32*
3. Medial and lateral plantar nerves.   *6.25*
4. From popliteal artery.   *6.24*

Page 121 A   Neck of fibula
       B   Tibialis posterior
       C   Peroneal artery
       D   Posterior tibial artery
       E   Tibial nerve
       F   Flexor digitorum longus
       G   Flexor hallucis longus
       H   Semimembranosus
       J   Semitendinosus
       K   Popliteal vein
       L   Tibial nerve

M   Biceps femoris
N   Medial head of gastrocnemius

1. Common peroneal nerve and its terminal branches.   *6.36*
2. Inversion of foot, weak plantarflexion of ankle.   *6.25*
3. Tibial part of sciatic nerve.   *6.20*
4. Head of fibula.   *6.20*
5. Tendo calcaneus.   *6.24*

Page 122 A   Fibrous flexor sheath
       B   Adductor hallucis (transverse head)
       C   Adductor hallucis (oblique head)
       D   Base of fifth metatarsal
       E   Tendon of abductor digiti minimi
       F   Abductor hallucis tendon
       G   Deep transverse metatarsal ligament
       H   Posterior part of abductor digiti minimi

J   Flexor accessorius
K   Abductor hallucis

1. Lateral plantar nerve.   *6.28*
2. Peroneus brevis tendon.   *6.36*
3. Plantar aponeurosis.   *6.26*
4. L5.   *6.6*

Page 123 A   Oblique popliteal ligament
       B   Semimembranosus tendon
       C   Fascia over popliteus, reinforced by semimembranosus insertion
       D   Popliteus tendon
       E   Meniscofemoral ligament
       F   Posterior cruciate ligament
       G   Lateral collateral ligament
       H   Medial head of gastrocnemius
       J   Anterior tibial artery
       K   Tibialis posterior

1. Popliteal vessels.   *6.21*
2. Rotates femur laterally during early stages of knee flexion.   *6.46*
3. Prevents anterior displacement of femur on tibia.   *6.46*
4. Becomes dorsalis pedis artery.   *6.36*

Page 124 A   Internal jugular vein
       B   Common carotid artery
       C   Arch of cricoid cartilage
       D   Phrenic nerve
       E   Scalenus anterior muscle
       F   Brachial plexus

G   Subclavius muscle
H   Thoracic duct
J   Inferior thyroid vein
K   Vagus nerve (X)
L   Brachiocephalic vein
M   Subclavian vein

Page 125 A Head of the mandible
B Inferior alveolar nerve
C Lateral pterygoid muscle
D Tensor palati muscle
E Levator palati muscle
F Posterior edge of middle concha
G Posterior edge of nasal septum
H Medial pterygoid muscle
J Uvula
K Posterior third of the tongue

1. Mandibular division of the trigeminal nerve (Vc). *7.22*
2. Pharyngeal plexus. *7.37*
3. Vomer. *7.32*
4. Elevation of the mandible. *7.22*
5. Glossopharyngeal nerve (IX) *7.28*

Page 126 A External jugular vein
B Sternocleidomastoid muscle
C Splenius capitis muscle
D Levator scapulae muscle
E Dorsal scapular nerve
F Trapezius muscle
G Scalenus posterior muscle
H Long thoracic nerve
J Inferior belly of omohyoid muscle
K Scalenus anterior muscle

1. Prevertebral layer of cervical fascia. *7.6*
2. Spinal part of accessory nerve (XI). *7.5*
3. Spinal part of accessory nerve (XI). *7.6*
4. Ansa cervicalis. *7.8*

Page 127 A Orbicularis oculi muscle
B Superficial temporal artery
C Zygomatic branch of facial nerve (VII)
D Parotid gland
E Parotid duct
F Masseter muscle
G Buccinator muscle
H Facial artery
J Orbicularis oris muscle
K Depressor labii inferioris muscle
L Platysma

1. Temporal and zygomatic branches of facial nerve (VII). *7.16*
2. External carotid artery. *7.19*
3. Ophthalmic division of trigeminal nerve (Va). *7.16*
4. Tympanic branch of IX→ lesser petrosal nerve→ otic ganglion→ auriculotemporal nerve. *7.19*
5. Mandibular division of trigeminal nerve (Vc). *7.20*

Page 128 A Anterior belly of digastric muscle
B Submandibular salivary gland
C Facial artery
D Jugulodigastric lymph node
E Anterior division of retromandibular vein
F Deep cervical lymph node
G Internal jugular vein
H Omohyoid muscle
J Supraclavicular lymph node
K Sternothyroid muscle
L Sternohyoid muscle

1. Mandibular division of trigeminal nerve (Vc). *7.25*
2. Jugular lymphatic trunk→ thoracic duct. *7.13*
3. Left brachiocephalic vein. *7.12*

Page 129 A Vault of skull
   B Temporalis muscle
   C Ethmoid air cell
   D Medial rectus muscle
   E Inferior concha
   F Buccinator muscle
   G Body of mandible
   H Anterior belly of digastric muscle
   J Genioglossus muscle
   K Sublingual salivary gland

L Mylohyoid muscle
M Masseter muscle

1. Into the middle meatus of
    the nose.   7.35
2. Inferior concha.   7.33
3. Facial nerve (VII).   7.16
4. Inferior alveolar nerve. 7.23
5. Submandibular duct or
    floor of mouth.   7.29

Page 130 A External carotid artery
   B Styloid process
   C Internal jugular vein
   D Posterior belly of digastric muscle
   E Accessory nerve (XI)
   F Vagus nerve (X)
   G Carotid sinus
   H Hypoglossal nerve (XII)
   J Stylohyoid muscle
   K Intermediate tendon of digastric
     muscle

1. Divides into superficial
    temporal and maxillary
    arteries.   7.12
2. Stylohyoid, styloglossus
    and stylopharyngeus. 7.25
3. Facial nerve (VII).   7.25
4. Inferior border of mandible
    near midline, via
    anterior belly.   7.25

Page 131 A Anterior cranial fossa
   B Olfactory bulb (I)
   C Optic nerve (II)
   D Anterior cerebral artery
   E Posterior cerebral artery
   F Midbrain
   G Tentorium cerebelli
   H Straight sinus

J Transverse sinus
K Superior sagittal sinus

1. Frontal lobes.   7.47
2. Optic foramen.   7.56
3. Internal carotid artery. 7.52
4. Cerebellum.   7.48
5. Internal jugular vein,
    via sigmoid sinus. 7.48

Page 132 A Deep temporal nerve
   B Lateral surface of lateral pterygoid
     plate
   C Maxillary artery
   D Medial pterygoid muscle
   E Lingual nerve
   F Inferior alveolar nerve
   G Chorda tympani
   H Middle meningeal artery
   J Auriculotemporal nerve
   K Facial nerve (VII)

1. Motor fibres to temporalis
    muscle.   7.22
2. Elevation of the mandible. 7.22
3. Lower teeth, alveolar ridge
    and skin over chin. 7.22
4. Taste and preganglionic
    parasympathetic. 7.23
5. Meninges inside skull. 7.50

Page 133 A Zygomatic arch
   B Incisive foramen
   C Palatine process of maxilla
   D Horizontal plate of palatine bone
   E Hamulus
   F Lateral pterygoid plate
   G Foramen lacerum
   H Jugular foramen

J Stylomastoid foramen
K Mastoid process

1. Nasopalatine nerve.   7.31
2. Tendon of tensor palati muscle. 7.37
3. Lateral pterygoid muscle. 7.22
4. Lymphatics only.   7.47
5. Facial nerve (VII).   7.19

Page 134 **A** Body of mandible
B Mylohyoid muscle
C Nerve to mylohyoid
D Styloglossus muscle
E Facial artery
F Lingual artery
G Hypoglossal nerve (XII)
H Hyoglossus muscle
J Submandibular duct
K Hyoid bone

1. Mandibular division of trigeminal nerve (Vc). *7.27*
2. Hypoglossal nerve (XII). *7.25*
3. At the anterior border of masseter. *7.12*
4. Depresses the tongue. *7.28*

Page 135 **A** Vagus nerve (X)
B Phrenic nerve
C Sympathetic trunk
D Inferior thyroid artery
E Scalenus anterior muscle
F Brachial plexus
G Vertebral artery
H Scalenus anterior tendon

J Thoracic duct
K Recurrent laryngeal nerve
L Subclavian artery
M Ansa subclavia
N Suprapleural membrane
O Internal thoracic artery

1. Pleural cavity and lung. *2.13*

Page 136 **A** Masseter muscle
B Ramus of mandible
C Medial pterygoid muscle
D Parotid gland
E Retromandibular vein
F Posterior belly of digastric muscle
G Vertebral artery
H Internal carotid artery
J Stylohyoid muscle
K Styloglossus muscle
L Stylopharyngeus muscle
M Internal jugular vein
N Accessory parotid gland
O Zygomatic branch of facial nerve (VII)
P Parotid duct
Q Masseter muscle
R Facial artery
S Platysma

T Cervical branch of facial nerve (VII)
U Parotid gland
V Sternocleidomastoid muscle

1. Elevation of the mandible. *7.20*
2. Into the vestibule of the mouth, opposite second upper molar. *7.18*
3. Unites with other vertebral artery to form basilar artery. *7.50*
4. Facial nerve (VII). *7.25*
5. Hypoglossal nerve (XII). *7.25*
6. Glossopharyngeal nerve (IX). *7.25*
7. Unites with subclavian vein to form brachiocephalic vein. *7.13*

Page 137 **A** Submandibular gland
B Hyoid bone
C Thyrohyoid muscle
D Superior thyroid vessels
E Sternothyroid muscle
F Sternocleidomastoid muscle
G Sternohyoid muscle
H Superior belly of omohyoid muscle
J Hyoid bone
K Sternohyoid muscle
L Ansa cervicalis
M Internal jugular vein
N Thoracic duct

O Inferior belly of omohyoid muscle
P Roots of brachial plexus
Q Superior thyroid vein
R Descending branch of hypoglossal nerve

1. C1 fibres in hypoglossal nerve. *7.8*
2. Turns head towards opposite shoulder. *7.5*
3. Suprascapular ligament and scapula. *7.8*
4. C1, C2 and C3. *7.9*
5. C1. *7.9*

Page 157 A Descending thoracic aorta
B Body of vertebra
C Right lung
D Transverse process
E Thoracic spinal cord
F Erector spinae
G Trapezius
H Latissimus dorsi

1. Midthoracic (T7) vertebra. *8.13*
2. Changes from left at T5
to anterior at T12
vertebra. *2.37*

Page 158 A Costal facet (demifacet, for rib
corresponding with vertebra)
B Inferior vertebral notch
C Spinous process
D Costal facet (entire)
E Facet on superior articular process
F Pedicle
G Lamina
H Body
I Transverse process
J Vertebral foramen

1. Second and tenth thoracic. *8.7*
2. Lower facet on head of
second rib. *8.7*
3. Active red marrow. *1.10*

Page 159 A Splenius
B Sternomastoid
C Serratus posterior superior
D Erector spinae
E External intercostal muscle
F Rib
G Serratus posterior inferior
H External abdominal oblique

1. Weak muscles of
ventilation. *8.12*
2. Anterior rami (intercostal
nerves). *8.12*